Beyond the Numbers: Foreign Direct Investment in the United States

CONTEMPORARY STUDIES IN ECONOMIC AND FINANCIAL ANALYSIS, VOLUME 83

Editors: Robert J. Thornton and J. Richard Aronson
Lehigh University

To Diana

Contemporary Studies in Economic and Financial Analysis
An International Series of Monographs

Edited by **Robert J. Thornton** and
J. Richard Aronson, *Lehigh University*

Volume 1 - **Hunt L.** - Dynamics of Forecasting Financial Cycles

Volume 2 - **Dreyer, J.S**. - Composite Reserve Assets in the International Money System

Volume 3 - **Altman, El.,** et al. - Application of Classification Techniques in Business, Banking and Finance

Volume 4 - **Sinkey, J.F.** - Problems and Failed Institutions in the Commercial Banking Industry

Volume 5 - **Ahlers, D.M.** - A New Look at Portfolio Management

Volume 6 - **Sciberras, E.** - Multinational Electronic Companies and National Economic Policies

Volume 7 - **Allen, L.** - Venezuelan Economic Developement: A Politico-Economic Analysis

Volume 8 - **Gladwin, T.M.** - Environment Planning and the Multinational Corporation

Volume 9 - **Kobrin, S.J.** - Foreign Direct Investment, Industrialization and Social Change

Volume 10 - **Oldfield, G.S.** - Implications of Regulation on Bank Expansion: A Simulation Analysis

Volume 11 - **Ahmad, J.** - Import Substitution, Trade and Development

Volume 12 - **Fewings, D.R.** - Corporate Growth and Common Stock Risk

Volume 13 - **Stapleton, R.C.** and **M.G. Subrahmanyam-** Capital Market Equillibrium and Corporate Financial Decisions

Volume 14 - **Bloch, E.** and **R.A. Schwartz**- Impending Changes for Securities Markets: What Role for the Exchange

Volume 15 - **Walker, W.F.** - Industrial Innovation and International Trading Performance

Volume 16 - **Thompson, J.K** - Financial Policy, Inflation and Economic Development: The Mexican Experience

Volume 17 - Omitted from Numbering

Volume 18 - **Hallwood, P.** - Stabilization of International Commodity Markets

Volume 19 - Omitted from Numbering

Volume 20 - **Parry, T.G.** - The Multinational Enterprise: International Investment and Host-Country Impacts

Volume 43 - **Plaut S.E.** - Import Dependence and Economic Vulnerability

Volume 44 - **Finn, F.J.** - Evaluation of the Internal Processes of Managed Funds

Volume 45 - **Moffitt, L.C.** - Strategic Management: Public Planning at the Local Level

Volume 46 - **Thornton, R. J., et al**. - Reindustrialization: Implications for U.S. Industrial Policy

Volume 47 - Omitted from Numbering

Volume 48 - **Von Pfeil, E.** - German Direct Investment in the United States

Volume 49 - **Dresch, S.** - Occupational Earnings,1967-1981: Return to Occupational Choice, Schooling and Physician Specialization

Volume 50 - **Siebert, H.** - Economics of the Resource-Exporting Country Intertemporal Theory of Supply and Trade

Volume 51 - **Walton, G.M.** - The National Economic Policies of Chile

Volume 52 - **Stanley, Q** - Managing External Issues: Theory and Practice

Volume 53 - **Natziger, E.W.** - Essays in Entrepreneurship, Equity and Economic Development

Volume 54 - Omitted from Numbering

Volume 55 - **Slottje, D.** and **J. Haslag.** - Macroeconomic Activity and Income Inequality in the United States

Volume 56 - **Thornton, R.F.** and **J.R. Aronson.** - Forging New Relationships Among Business, Labor and Government

Volume 57 - **Putterman, L.** - Peasants, Collectives and Choice: Economic Theory and Tanzanian Villages

Volume 58 - **Baer, W.** and **J.F. Due** - Brazil and the Ivory Coast: The Impact of International Lending, Investment and Aid

Volume 59 - **Emanuel, H., et al.** - Disability Benefits: Factors Determining Applications and Awards

Volume 60 - **Rose, A.** - Forecasting Natural Gas Demand in a Changing World

Volume 61 - Omitted from Numbering

Volume 62 - **Canto V.A.** and **J.K Dietrich** - Industrial Policy and Industrial Trade

Volume 63 - **Kimura, Y.** - The Japanese Semiconductor Industry

Volume 64 - **Thornton, R., Hyclak, T.** and **J.R. Aronson** - Canada at the Crossroads: Essays on Canadian Political Economy

Volume 65 - **Hausker, A. J.** - Fundamentals of Public Credit Analysis

Volume 66 - **Looney, R.** - Economic Development in Saudi Arabia: Consequences of the Oil Price Decline

Volume 67 - **Askari, H.** - Saudi Arabias Economy: Oil and the Search for Economic Development

Volume 21 - **Roxburgh, M.** - Policy Responses to Resource Depletion: The Case of Mercury

Volume 22 - **Levich, R.M.** - The International Money Market: An Assessment of Forecasting Techniques and Market Efficiency

Volume 23 - **Newfarmer, R.** - Transnational Conglomerates and the Economics of Dependent Development

Volume 24 - **Siebert, H.** and **A.B. Anatal** - The Political Economy of Environmental Protection

Volume 25 - **Ramsey, J.B.** - Bidding and Oil Leases

Volume 26 - **Ramsey J.B.** - The Economics of Exploration for Energy Resources

Volume 27 - **Ghatak, S.** - Technology Transfer to Developing Countries: The Case of the Fertilizer Industry

Volume 28 - **Pastre, O.** - Multinationals: Bank and Corporation Relationships

Volume 29 - **Bierwag, G.O**. - The Primary Market for Municipal Debt: Bidding Rules and Cost of Long Term Borrowing

Volume 30 - **Kaufman, G.G**. - Efficiency in the Municipal Bond Market: The Use of Tax Exempt Financing for "Private" Purposes

Volume 31 - **West, P.** - Foreign Investment and Technology Transfer: The Tire Industry in Latin America

Volume 32 - **Uri, N.D.** - Dimensions of Energy Economics

Volume 33 - **Balabkins, N.** - Indigenization and Economic Development: The Nigerian Experience

Volume 34 - **Oh, J.S**. - International Financial Management: Problem, Issues and Experience

Volume 35 - **Offlcer, L.H.** - Purchasing Power Parity and Exchange Rates: Theory Evidence and Relevance

Volume 36 - **Horwitz, B.** and **R. Kolodny** - Financial Reporting Rules and Corporate Decisions: A Study of Public Policy

Volume 37 - **Thorelil, H.B.** and **G.D. Sentell** - Consumer Emancipation and Economic Development

Volume 38 - **Wood, W.C.** - Insuring Nuclear Power: Liability, Safety and Economic Efficiency

Volume 39 - **Uri, N.D.** - New Dimensions in Public Utility Pricing

Volume 40 - **Cross, S.M**. - Economic Decisions Under Inflation: The Impact of Accounting Measurement Error

Volume 41 - **Kautman, G.G**. - Innovations in Bond Portfolio Management: Immunization and Duration Analysis

Volume 42 - **Horvitz, P.M.** and **R.R. Pettit** - Small Business Finance

Volume 68 - **Godfrey, R.** - Risk Based Capital Charges for Municipal Bonds

Volume 69 - **Guerard, J.B** and **M.N. Gultekin** - Handbook of Security Analyst Forecasting and Asset Allocation

Volume 70 - **Looney, R.E** - Industrial Development and Diversification of the Arabian Gulf Economies

Volume 71 - **Basmann, R.L., et al.** - Some New Methods for Measuring and Describing Economic Inequality

Volume 72 - **Looney, R.E.** - The Economics of Third World Defense Expenditures

Volume 73 - **Thornton, R.J.** and **A.P. O'Brien** - The Economic Consequences of American Education

Volume 74 - **Gaughan, P.A.** and **R.J. Thornton** - Litigation Economics

Volume 75 - **Nafzinger, E. Wayne** - Poverty and Wealth: Comparing Afro-Asian Development

Volume 76 - **Frankurter, George** and **Herbert E. Phillips** - Forty Years of Normative Portfolio Theory: Issues, Controversies, and Misconceptions

Volume 77 - **Kuark, John Y.T**. - Comparative Asian Economics

Volume 78 - **Stephenson, Kevin** - Social Security: Time for a Change

Volume 79 - **Duleep, Harriet Orcutt** and **Wunnava, Panindra V.** - Immigrants and Immigration Policy: Individual Skills, Family Ties, and Group Identities

Volume 80 - **Frankfuter, G.M.** and **McGoun, E.G.** - Toward Finance with Meaning The Methodology of Finance: What It Is and What It Can Be

Volume 81 - **Askari, H., Nowshirvani, V.** and **Mohamed Jaber -** Economic Development in the GCC: The Blessing and the Curse of Oil

Volume 82 - **Moroney, John R. -** Exploration, Development, and Production: Texas Oil and Gas, 1970-1995

Beyond the Numbers: Foreign Direct Investment in the United States

by WILLIAM L. CASEY, JR.
Department of Economics
Babson College

Greenwich, Connecticut *London, England*

Casey, William L.
Beyond the numbers: foreign direct investment in the United States/by William L. Casey. Jr.
p. cm. — (Contemporary studies in economic and financial analysis: v. 83)
Includes index.
ISBN 0-7623-0383-2
1. Investment, Foreign—United States. 2. Investments, Japanese—United States. I. Title. II. Series.
HG4910,C358 1998
332.57'3'0973—dc21 98–5306
CIP

55 Old Post Road, No. 2
Greenwich, Connecticut 06830

JAI PRESS LTD.
38 Tavistock Street
Covent Garden
London WC2E 7PB
England

ISBN: 0-7623-0383-2

Library of Congress Catalog Card Number: 98-5306

Manufactured in the United States of America

CONTENTS

Preface and Acknowledgments

During the 1990s, there has been a proliferation of articles and reports both in the commercial press and in academic journals examining the phenomenon of economic globalization. The freer movements of goods, services, labor and capital across international boundaries, spurred by technological change and the dismantling of political barriers, are raising new issues in the arena of public policy and are mandating new solutions to old problems. For example, those central banks today that base monetary policy on the looseness or tightness of domestic labor markets alone render a disservice. In the newly emerging world economy, resource market conditions abroad affect resource market flexibility at home, and public policy makers must look beyond artificial national boundaries in crafting effective policies of adjustment.

The focus of this book is on one aspect of economic globalization, namely, the accelerating flow of foreign direct investment across national boundaries. Historically, nation states have promoted economic specialization, while adhering to the principle of comparative advantage, primarily through the trading of goods and services. Although world trade has certainly expanded over the past two decades, foreign direct investment has accelerated at a much faster pace to the extent that, by 1995, the global sales of the foreign affiliates of multinational corporations had grown to exceed the total market value of world trade in goods and services (World Trade Organization, Press Release, October 9, 1996).

Predictably, the United States has been at the center of this global phenomenon, emerging over the past two decades as the world's largest exporter and importer of foreign direct investment capital. The extensive overseas investments and operations of U.S. multinational corporations have attracted considerable attention in the economic literature. The growing presence of foreign multinationals on U.S. soil has attracted less. It is the purpose of the book to shed light on issues and questions relating to the latter by identifying the factors and conditions that have motivated foreign companies to enter the U.S. market as producers, rather than opting to serve the market as exporters or as licensors.

Foreign direct investment in the United States (FDIUS) throughout the post-World War II period is examined in the book, although special attention is directed at the accelerating pace of inward FDI during the 1980s and 1990s. The analysis is designed not only to identify specific factors or conditions that have attracted capital inflows, but also to indicate how foreign investment motivations have changed over time in response to structural changes in the United States and global economies.

This book is the final product of a series of research projects that I conducted over the past three years. A major part of this effort was supported by Babson College's Board of Research. The support came in the form of a grant, released time from part of my teaching responsibility, ongoing encouragement and a steady stream of good advice. The board, under the leadership of Professors Nan Langowitz, C. J. McNair, and Richard Flanagan, has made significant contributions to the research productivity of the Babson faculty in recent years and I have benefited from the wisdom and dedication of board members. I owe no greater debt of thanks than to Professor McNair who assisted me in the editing, publishing and marketing of the book. In my view, she epitomizes the highest level of integrity, professionalism and collegiality in the world of college faculty research.

Several members of the Economics Department at Babson College were generous with their time and insightful with their advice in reading earlier versions of the manuscript or in suggesting references. I would like to thank in particular John Marthinsen, Lidija Polutnik, Kostas Axarloglou, Mark Tomass and Kent Jones. The library staffs at Babson and Wellesley Colleges were both helpful in locating documents and in securing copies of key references for me. During the early drafting of the book, administrative turnover in the Economics Division produced some word processing chaos. Fortunately, Lynn Fennell arrived in 1996 and rectified the situation with her patience, diligence, dedication and word processing skills.

Outside of Babson, I would like to thank Professor John Stevens of Franklin College of Indiana who co-authored a paper with me years ago that served as one of the cornerstones of Chapter III. Also, I would like to thank the Japan Society of New York as well as numerous officials from the economic development agencies of the 50 U.S. states for their data and information. Earlier versions of Chapters II, III and IV were presented as conference papers and I appreciate the feedback and advice from attending scholars associated with the Academy of International Business, the Eastern Economic Association and the Academy of Business Administration.

To all of these people, I acknowledge their indispensable assistance and offer my sincere thanks.

William Casey
Department of Economics
Babson College

Chapter I

Introduction

In 1967, French journalist J. J. Servan-Schreiber authored a best-selling book, *Le Defi Americain* (*The American Challenge*), in which he predicted that, within fifteen years of the publication date, the three major superpowers in the world would likely be: (1) the United States, (2) the Soviet Union and (3) U.S. business interests in Europe.[1] Servan-Schreiber based his forecast on a phenomenon that had been causing concern and even alarm throughout Western Europe during the 1950s and 1960s, namely, the growing inclinations on the part of U.S. companies to move to Europe as producers in order to serve the market as foreign direct investors, rather than as exporters or licensors.

Servan-Schreiber's prediction was not accurate for several reasons. First of all, the pace and direction of the outward flow of U.S. foreign direct investment changed in the 1970s and beyond. U.S. multinationals, while still motivated to invest in Europe, broadened their investment horizons worldwide. Second, with post-common market European economic growth, the prospects for Europe becoming an economic satellite to the United States became progressively more remote, despite the expanding presence of U.S. multinationals on the continent. Third, and perhaps most importantly, the possibility of lopsided U.S. foreign direct investment producing economic and then political domination in Europe dissolved as European and other multinationals became more active and aggressive in the late 1960s and beyond in acquiring productive property assets in the United States. The foreign acquisition of a nation's productive resources can be alarming with significant political, social and cultural consequences. However, the impact is less disconcerting if the flow of foreign direct investment between countries or regions is two-way, not one-way, producing an international sharing of property assets.

The purpose of this book is to examine in depth the factors that have motivated foreign firms to come to the United States in recent decades as producers in larger and larger numbers and at an accelerating pace, thereby making foreign direct investment a two-way street for the United States. The first step is to establish the importance of the phenomenon and to weigh the benefits and the costs.

BEYOND THE NUMBERS: WHAT ARE THE ISSUES?

The heavy infusion of FDI capital into the United States in recent decades has been part of a broader trend towards the greater participation of foreign multinationals from all regions in the global economy (Graham and Krugman 1995, p. 1). Although the literature on the impact of recent FDIUS does not contain the passion of Servan-Schreiber (1967) in his warnings of U.S. dominance of Europe through earlier lopsided USFDI, there are significant issues today. With broader acceptance of the international sharing of property assets, the rhetoric used in debating the issues has been toned down[2] and there is more of a tendency to weigh the costs and risks in relation to benefits. Although the central purpose of this study is to examine the root causes of FDIUS, not the effects, a brief summary of the latter may be useful in demonstrating the importance of the former.

The Dark Side of FDI: Potential Costs and Risks

Potentially, there are both economic and non-economic costs and concerns that may arise if a country opens its borders to FDI. On the economic side, local firms may argue that the presence of foreign multinationals translates into "unfair" competition; labor unions may worry about employment and real wage effects and government officials in target countries may agonize over several economic issues, including the potential for adverse balance-of-payment effects, the possibility that a depreciating currency may permit foreign firms to buy property assets in the target country at fire sale prices, the danger that technological secrets may be transferred abroad through FDI and even that FDI may put domestic firms at strategic disadvantages vis-a-vis foreign multinationals. Specifically, concerns may arise that subsidiaries of multinationals with superior resources and strategic advantages may outgun local companies and, in the process: (1) replace good jobs with bad jobs, (2) lower real wages, (3) transfer jobs back home because of a high import propensity, (4) cause deterioration in the host country's trade balance, (5) transfer technology back home compromising the host country's technological secrets and (6) time property acquisition in the host country to take advantage of fire-sale opportunities presented by exchange rate distortions.[3]

On the non-economic side, an examination of the potential dark side of FDI would include the cultural and social tensions that may arise with local labor working for foreign management, the political risks of undue lobbying influence exercised by foreign multinationals over local public policy as well as the danger that foreign ownership of certain property assets and industries in host countries may compromise national security.

Interestingly, despite the proliferations of warnings about the economic, social and political risks of the heavy infusion of FDI capital into the United States over the past three decades (particularly in the popular press), very little hard evidence has surfaced in support of the negative effects cited above.[4] For example, one of the central purposes of the highly regarded Graham and Krugman study (1995) was to examine the economic impact and political effects of FDIUS. Despite the fact that Graham and Krugman found some problems in the areas of U.S. national security and identified some valid logic behind other political concerns, they concluded that most economic concerns "about harmful foreign firm behavior are not borne out by the experience with FDI so far" (p. 4).

Nevertheless, public opinion which influences public policy may arise with or without scientific support. FDIUS remains an issue of concern in the minds of many Americans, apart from the evidence uncovered by scholars.

The Positive Side of FDI: Potential Benefits and Gains

Gains from foreign direct investments, as in the case of gains from international trade, are rooted in traditional microeconomic analysis. In a world of free capital flows, countries (including FDI hosts) benefit from the international sharing of productive property resources in several ways. First of all, economic efficiency is promoted through increased competition as multinational firm operations spread globally serving to dissipate local monopoly power. Consumer welfare obviously increases in host countries if multinationals are permitted to move into market areas previously dominated by local oligopolies or monopolies, which were protected by entry barriers in the form of international capital restrictions.

Also, as barriers break down, firms are able to expand operations overseas therefore capturing economies of scale and scope. In riding down its long run cost curve, the globally expanding multinational is able to cut costs, and increased competition in a world without capital walls would then force the firm to pass on at least part of its cost savings to host country consumers in the form of lower prices. Furthermore, to the extent that FDI facilitates trade in goods and services, multinationals promote comparative advantage. Inasmuch as countries have different resource endowments, FDI induced trade "enables them to specialize and benefit from their differences" (Graham and Krugman 1995, p. 57).

Less measurable than the above, but potentially more important, is the FDI induced flow of knowledge across international borders. Benefits to host countries accrue in the form of positive spillovers as multinationals bring to their overseas operations their up-to-date technology, management skills, marketing techniques, and so on. This highlights one of the significant differences between foreign portfolio investment and foreign direct investment. With the former, the host country imports only capital, whereas, with the latter, much more is imported particularly in the form of intangible spillovers.

As in the case of the potential costs of FDI, there are also areas of controversy and disagreement with reference to potential benefits or gains. Not surprisingly, these tend to be more country specific. For example, in the popular debate over the relative costs and benefits of FDI in the United States, proponents focus a disproportionate amount of attention on the job creating potential.

According to the conventional wisdom, multinationals expand employment. However, Graham and Krugman (1995, p. 61) argue that FDI has essentially no long-term net impact on aggregate employment at the national level because, over the long term, employment is supply, not demand, determined. On the other hand, it is generally held that FDI does have a regional distribution effect on employment if, for example, certain state governments successfully attract FDI with investment incentive programs and others do not. Interestingly, if these two positions are valid, incentive programs at the state/regional level would be justified, but not at the national level.[5]

Other controversial areas include the "balance of trade" effects and the "net savings" effects of FDI in the host country. Do multinationals have a greater propensity to import than to export? Does FDI contribute positively or negatively to host countries' balance of trade and current account positions over time? Do FDI capital flows contribute positively to host countries' growth performances by supplementing deficiencies in domestically generated savings or are long-term growth effects negative because of an unhealthy dependency on the part of host countries in externally generated funds? Is growth retarded as multinationals ultimately repatriate earnings in excess of new gross capital inflows? Does FDI create external debt problems with unhappy balance of payments consequences?

Clearly, these are questions of importance with significant market and public policy implications and ramifications. They are cited here not with the promise to provide definitive answers but rather to point out what is at stake as well as to demonstrate the need for further study of the causes and consequences of the phenominal growth of foreign direct investment in the United States in recent decades.

THE FOCUS AND ORGANIZATION OF THE BOOK

As indicated earlier, the focus of this book is not on the effects or consequences of foreign direct investment; rather, it is on the causes. The principal objective is to shed light on investment motivation, that is, the determinants of FDIUS. What market factors or public initiatives (in either the host or investing country) motivate the foreign firm to serve the U.S. market internally as a producer, rather than an exporter or licensor?

Chapter II reviews the development and evolution of foreign direct theory over the past four decades, ranging from early attempts by economists to build theory on a traditional microeconomic foundation to more recent interdisciplinary efforts to identify the strategic dimensions of multinational firm activities. The chapter is both descriptive and analytical with an attempt to develop a framework of analysis within which the validity and comprehensiveness of the specific theories can be evaluated.

Of course, the ultimate test of the validity of a theory is whether the expressed relationship truly describes reality. Is the theory empirically verifiable? It is the purpose of Chapter III to present and examine the empirical evidence relating to the determinants of FDIUS. Evidence is drawn from the findings of case studies, surveys, cross-sectional and time-series statistical analyses. In the identification and examination of FDIUS motivational factors, the analysis in the chapter is divided into two distinct time periods. The determinants of FDIUS during the 1960s and 1970s are compared and contrasted with those factors or conditions governing FDI decision making over the past two decades. Changes over time in the decision making attitudes of multinational corporate (MNC) management are noted and examined.

Given the dramatic rise in Japanese FDIUS in the 1970s and 1980s and the more recent retreat, special factors seem to govern the investment decision making of Japanese multinationals. It is the purpose of Chapter IV to identify and examine those special factors that have both "pushed" and "pulled" Japanese FDI to U.S. shores and to evaluate to what extent and in what ways they are unique.

The central focus of Chapters III and IV is the identification of those market factors that have influenced the FDI decisions of foreign multinationals in general and of Japanese multinationals in particular.[6] The focus shifts in Chapters V and VI to public sector policies and initiatives that influence the FDI decision making process. In Chapter V, the policies and programs of the Federal governments are examined by: (1) distinguishing those which are explicit from those which are implicit and (2) distinguishing those that offer FDI incentives from those which produce disincentives. The focus here, of course, is on recent and current U.S. government policy on inward FDI.

In Chapter VI, state government policy on FDI is scrutinized with special attention paid to the aggressive competition among states, which has intensified over the past three decades. State-sponsored investment incentive programs are identified and analyzed. Then, an attempt is made to evaluate the effectiveness of incentive packages by matching those states which historically have been most successful in attracting FDI with those that have been most aggressive in offering incentives to potential foreign investors. Both primary and secondary data are used in conducting the analysis.

The final chapter of the book summarizes the major findings and conclusions of the study, examining the implications and ramifications of the same and ends with the identification of those important issues that remain unresolved, thereby calling for further research and study.

Before this comprehensive examination of FDIUS takes place in subsequent chapters, it is necessary to develop a statistical overview. The next section examines the quantitative dimension of this phenomenon which has transformed the U.S. over a short period of time into the world's largest recipient of FDI, matching the country's status as the world's largest source (OECD 1995, p. 11).

STATISTICAL SUMMARY OF FOREIGN DIRECT INVESTMENT IN THE UNITED STATES

A foreign direct investment entails a controlling interest in the foreign investment property as opposed to a portfolio or minority shareholding. The OECD is credited with the most widely accepted definition of FDI, namely, as foreign control of 10 percent or more of the ordinary shares or voting power of a foreign enterprise or an effective voice in the management of the same (Economic Policy Council 1991).

Foreign direct investment is measured as earnings retained by operating subsidiaries of foreign companies and the transfer of funds from parent organizations in investing countries to subsidiaries or branches in host countries. Both debt and equity raised capital are included as transfers. When capital is borrowed by the foreign subsidiary either within the host country or from capital markets, such borrowings are not included.

One measure, used below, in examining the quantitative dimensions of foreign direct investment flowing into and out of the United States is the annual flow of FDI collected and organized by the Bureau of Economic Analysis (BEA) of the U.S. Department of Commerce. Inflow data are available both by country of origin and by sector. A second measure, also collected by the BEA and used below, is the stock of FDI accumulated over time. It is the sum of foreign owners' equity on the balance sheets of foreign affiliates, plus the sum of net lending of these affiliates from parent companies. This statistical series is particularly useful in measuring the relative importance of FDI inflows and is also available by country of origin and by sector.

The Quantitative Dimensions: FDIUS by Country of Origin

Table I.1 shows the reported current dollar value of total foreign direct investment in the United States (net capital inflows) from particular areas for selected years (1950-1995). Table I.2 reveals the accumulated stock of FDI in the United States for the same areas over the same time span with the addition of preliminary figures for 1996. Several broad generalizations emerge from the data.

First, both tables reveal that net capital inflows were relatively low in the 1950s and 1960s and, although accumulated stock roughly tripled from the late 1950s to 1970, these were increases from a very low base. The United States attracted very modest levels of FDI prior to the late 1960s. The data also

Table I.1. Foreign Direct Investment in the United States: Capital Inflows by Region of Origin End of Year; Outflows (–) (in million of dollars)

	All areas	*Canada*	*Europe*	*Japan*
1950	80	37 (46.2)	41 (51.3)	N/A (–)
1954	124	42 (33.9)	79 (63.7)	N/A (–)
1958	98	31 (31.6)	46 (46.9)	N/A (–)
1962	132	43 (32.6)	62 (47.0)	24 (18.1)
1966	86	2 (2.3)	90 (104.7)	−24 (–)
1970	1,030	238 (23.1)	730 (70.9)	−1 (–)
1974	2,745	540 (20.0)	1,690 (61.6)	159 (6.0)
1978	7,874	530 (6.7)	5,424 (68.9)	994 (12.6)
1982	13,842	−1439 (–)	10,610 (76.7)	1,987 (14.3)
1986	34,091	2,547 (7.5)	21,730 (63.7)	7,268 (21.3)
1988	59,424	1,179 (2.0)	32,996 (55.5)	17,287 (29.1)
1990	37,213	13 (0.1)	16,314 (43.8)	17,336 (46.6)
1991	22,799	103 (0.5)	12,013 (52.7)	12,782 (56.1)
1992	18,885	1,335 (7.1)	8,105 (42.9)	5,871 (31.1)
1993	41,738	3,103 (7.4)	34,009 (81.5)	61 (0.2)
1994	50,060	3,968 (7.9)	31.441 (62.8)	6.442 (12.9)
1995	60,848	4,489 (7.3)	51,793 (85.1)	5,252 (8.6)

Notes: Percentages of total are in parenthesis. For some years the Percentages above may exceed 100% because of capital outflows (–) for other countries.

Source: *Survey of Current Business,* various issues.

Table I.2. Foreign Direct Investment in the United States: Position by Region of Origin at End of Year (in million of dollars)

	All areas	*Canada*	*Europe*	*Japan*
1950	3,391	1,029 (30.3)	2,227 (65.7)	N/A (–)
1954	3,981	1,188 (29.8)	2,366 (59.9)	N/A (–)
1958	4,940	1,631 (33.0)	3,080 (62.3)	N/A (–)
1962	7,612	2,064 (27.1)	5,245 (68.9)	112 (1.5)
1966	9,054	2,439 (26.9)	5,273 (58.2)	103 (1.1)
1970	13,270	3,117 (23.5)	9,554 (72.0)	229 (1.7)
1974	22,421	4,930 (22.0)	14,629 (65.2)	504 (2.2)
1978	40,831	6,166 (15.1)	27,895 (68.3)	2,688 (6.6)
1982	123,590	11,435 (9.3)	82,767 (67.0)	9,697 (7.8)
1986	220,414	20,318 (9.2)	144,181 (65.4)	26,824 (12.2)
1988	328,850	27,361 (8.3)	216,418 (65.8)	53,354 (16.2)
1990	403,735	27,733 (6.9)	256,496 (63.5)	83,498 (20.7)
1991	419,108	36,835 (8.8)	256,053 (61.1)	95,142 (22.7)
1992	427,566	37,843 (8.9)	255,570 (60.0)	99,628 (23.3)
1993	464,110	40.143 (8.7)	287,084 (61.9)	99,208 (21.4)
1994	504,401	43,223 (8.6)	312,876 (62.0)	103,120 (20.4)
1995	560,088	46,005 (8.2)	360,762 (64.4)	108,582 (19.4)
1996	630,045	53,845 (8.5)	410,425 (65.1)	118,116 (18.7)

Notes: Percentages of total are in parenthesis. Also, 1996 data are preliminary.

Source: *Survey of Current Business,* various issues.

reveal that during the early decades of the post-World War II period, investment capital came predominantly from Canada and Europe. Indeed, during the 1950s, roughly two-thirds of accumulated FDI stock originated from Europe and one-third from Canada. Investments from other countries or regions were typically recorded by BEA in a catchall (others) category.

The decade of the 1970s witnessed dramatic change in both the level and origin of FDIUS. Net capital inflows in the aggregate rose sharply (Table I.1), although part of this rise in current dollar values was attributable to the substantial inflation of the period. The significant rise in inward FDI during this period increased the stock of accumulated capital from $13.3 billion in 1970 to $123.6 billion in 1982 (Table I.2).

As important as the aggregate rise in inward FDI during the 1970s was the emergence of Japan as a major direct investor in the global economy. Relegated to the "others" category in the 1950s, Japan's relative contribution to the accumulated stock of FDI capital in the United States averaged less than 2 percent in the 1960s and early 1970s. However, by 1982, Japan's relative share had risen to approximately 8 percent (Table I.2). The surge in Japanese FDIUS had begun. Interestingly, this phenomenon coincided with the decline in the relative importance of Canadian FDIUS (Tables I.1 and I.2). Net capital inflows from Canada rose absolutely in the 1970s but not as sharply as investments from Europe and Japan. Canada's share of total FDI stock in the United States, which averaged one-third of the total in the 1950s and 1960s, fell to under 10 percent by the early 1980s. Finally, European FDIUS certainly increased in absolute value during the 1970s, but the region's relative share remained fairly constant at the "two-thirds of total" level (Table I.2).

The decade of the 1980s witnessed the most dramatic surge in FDIUS historically, especially after 1985. During this period the United States became the world's largest recipient of inward FDI, while remaining the largest source of the stock of outward FDI (Rutter 1991, p. 20). Annual capital inflows rose from $7.9 billion in 1978 to over $59.4 billion in 1988 (Table I.1). Accumulated stock over this same ten year period surged from $40.8 billion to $328.9 billion (Table I.2). This FDI "invasion" of the United States was led by Japan, which was an insignificant contributor as recently as the mid-1970s. Japan's relative share of accumulated FDI stock in the United States rose dramatically from 6.6 percent in 1978 to 20.7 percent by 1990. Canadian and European investments increased sharply in absolute numbers, but their relative shares in total stock dropped slightly (Table I.2). Six countries (Japan, Great Britain, Germany, the Netherlands, France and Canada) dominated the FDIUS scene during the 1980s, accounting for nearly 90 percent of the rise in the FDIUS position during the mid- and late 1980s (Borghese 1993, p. 23).

Because of recessionary conditions in the United States and elsewhere, FDIUS continued to rise in the early 1990s but at a decidedly slower pace (Tables I.1 and I.2). The slowdown in 1990 and 1991 was widespread across virtually all major source nations (Borghese 1993, p. 18). In reference to relative shares, the most significant retrenchment came from Japan. Weakened by internal structural and cyclical problems and troubled by unanticipated losses on existing investments in the United States, particularly in real estate, Japanese multinationals retreated. Despite a spark of renewed investment interest in 1994, 1995 and 1996, Japan has been investing in the mid-1990s well below the levels reached in the mid- and late 1980s (Tables I.1 and I.2). Interestingly, the most recent data reveals that the Japanese retrenchments have been largely offset by a mid-1990s surge in European FDIUS. Intuitively, one might surmise that recent public policy turmoil, labor market rigidity, and other structural problems in Europe have enhanced the relative appeal of the U.S. market for European multinationals.

Quantitative Dimensions: FDIUS by Sector

Tables I.3 and I.4 disaggregate FDIUS annual capital inflows and accumulated stocks respectively for selected years (1950-1994). During the 1950s and early 1960s the relatively small levels of inward FDI, noted earlier, were fairly evenly distributed among the manufacturing, petroleum (not shown because of declining importance over time) and financial (including banking and insurance) sectors. Analysis of early trends in sec-

Table I.3. Foreign Direct Investment in the United States: Capital Inflows by Sector at End of Year; Outflows (–) (in million of dollars)

	Total	*Manufacturing*	*Wholesale and Retail Trade*	*Banking/ Finance*	*Insurance*	*Real Estate*
1950	80	31 (38.8)	N/A (–)	(a)	39 (48.8)	N/A (–)
1954	124	56 (45.2)	N/A (–)	(b)	39 (31.5)	N/A (–)
1958	98	70 (71.4)	N/A (–)	(c)	19 (19.4)	N/A (–)
1962	132	41 (31.0)	N/A (–)	11 (8.3)	36 (27.2)	N/A (–)
1966	186	111 (59.7)	−39 (–)	13 (7.0)	64 (34.4)	N/A (–)
1970	1,030	545 (59.9)	−19 (–)	15 (1.5)	44 (4.3)	N/A (–)
1974	2,745	1,348 (49.1)	230 (8.3)	222 (8.1)	93 (3.4)	N/A (–)
1978	7,897	3,172 (40.2)	1,924 (24.4)	620 (7.9)	455 (5.7)	N/A (–)
1982	13,842	2,740 (19.8)	2,936 (21.2)	2,324 (16.8)	805 (5.8)	2,500 (18.1)
1986	34,091	11,865 (34.8)	6,438 (18.9)	3,653 (10.7)	3.702 (3.1)	3,099 (9.1)
1988	59,424	33,138 (55.8)	7,490 (12.6)	3,542 (6.0)	1,599 (2.7)	3,469 (5.8)
1990	37,213	11,610 (31.2)	7,778 (20.9)	−2,792 (–)	4,430 (11.9)	4,764 (12.8)
1991	22,799	7,286 (32.0)	3,831 (16.8)	5,896 (25.9)	4,442 (19.5)	−256 (–)
1992	18,885	7,533 (39.9)	2,253 (11.9)	5,902 (31.2)	1,780 (23.6)	−295 (–)
1993	41,738	13,905 (33.3)	4,008 (9.6)	22,250 (53.3)	1,874 (4.4)	−491 (–)
1994	50,060	21,694 (43,3)	8,748 (17.4)	5,388 (10.8)	2,789 (5.6)	438 (0.8)
1995	60,848	26,246 (43.1)	5,877 (9.7)	15,979 (26.3)	4,057 (6.7)	−1,199 (–)

Notes: [a] Suppressed to avoid disclosure of data of industrial companies.

Source: U.S. Department of Commerce, *Survey of Current Business*, various issues.

Table I.4. Foreign Direct Investment in the United States: Position by Sector at End of Year; Outflows (–) (in million of dollars)

	Total	*Manufacturing*	*Wholesale and Retail Trade*	*Banking/ Finance*	*Insurance*	*Real Estate*
1950	3,391	1,138 (33.6)	N/A (–)	(a)	1,065 (31.4)	N/A (–)
1954	3,981	1,251 (31.4)	N/A (–)	(b)	1,211 (30.4)	N/A (–)
1958	4,940	1,860 (37.7)	N/A (–)	(c)	1,492(30.2)	N/A (–)
1962	7,612	2,885(37.9)	750 (9.8)	N/A (–)	1,943 (25.5)	N/A (–)
1966	9,054	3,789 (41.9)	739 (8.2)	N/A (–)	2,072 (22.9)	N/A (–)
1970	13,270	6,140 (46.3)	994 (7.5)	N/A (–)	2,256 (17.0)	N/A (–)
1974	22,421	10,387 (46.3)	4,387 (19.6)	N/A (–)	1,289 (5.8)	N/A (–)
1978	40,831	17,202 (42.1)	9,161 (22.4)	N/A (–)	2,773 (6.8)	N/A (–)
1982	123,590	44,100 (35.7)	22,656 (18.3)	10,184 (8.2)	7,772 (6.2)	11,397 (9.2)
1986	220,414	71,963 (32.6)	42,920 (19.5)	19,633 (8.9)	15,345 (7.0)	22,512 (10.2)
1988	328,850	121,434 (36.9)	64,930 (19.7)	19,577 (6.0)	20,252 (6.2)	31,929 (9.7)
1990	403,735	159,998 (39.6)	61,996 (15.4)	32,164 (8.0)	26,273 (6.5)	34,626 (8.6)
1991	419,108	157,115 (37.5)	65,334 (15.6)	36,283 (8.7)	33,341 (8.0)	33,577 (8.0)
1992	427,566	158,873 (37.2)	67,888 (15.9)	41,313 (9.7)	35,834 (8.4)	32,406 (7.6)
1993	464,110	166,397 (35.9)	72,805 (15.7)	67,900 (14.6)	40,376 (8.7)	28,391 (6.1)
1994	504,401	184,484 (36.6)	79,542 (15.8)	71,412 (14.2)	41,370 (8.1)	28,389 (5.6)
1995	560,088	210,312 (37.5)	85,086 (15.2)	89,126 (15.9)	47,283 (8.4)	26,518 (4.7)
1996	630,045	234,323 (37.2)	92,945 (14.8)	102.088 (16.2)	59,566 (9.5)	30,118 (4.8)

Notes: [a] Prior to 1960, banking and finance totals were combined with insurance. Percentages of total are in parenthesis. Also, 1996 data are preliminary.

Source: *Survey of Current Business*, various issues.

tor by sector investments is not easy, however, because of the paucity of disaggregated BEA data.

It is clear from the data that the relative importance of FDIUS in manufacturing grew during the late 1960s and early 1970s (Tables I.3 and I.4). By the end of this period, over 45 percent of total FDIUS capital stock was in the manufacturing sector. Foreign investments in manufacturing in relation to total investments eroded slightly in the mid- and late 1970s, but remained in excess of 40 percent throughout this period. FDIUS in manufacturing continued to increase in absolute terms during the early and mid-1980s, but its relative share in the total stock dropped sharply to less than one-third by 1986 (Table I.4).

In effect, the dramatic surge in FDIUS that gained momentum in the 1970s and took off in the 1980s was not led by investments in manufacturing. Faster rates of growth were recorded by investments in wholesale and retail trade (1970s and 1980s) and by the combined package of financial services (including banking and insurance) during the 1980s (Tables I.3 and I.4). Most importantly, manufacturing investments in the early and mid-1980s failed to keep pace with the surge in real estate investments coming mostly from Japan (Borghese 1993; Rutter 1991).

One should not overstate the importance of this sectorial shift in FDIUS, however. Despite the aforementioned erosion in the relative share of foreign investments in manufacturing during the 1970s and early to mid-1980s, this sector consistently captured the largest share (compared to any one other sector) of total FDIUS throughout the time period under study. Furthermore, although the pace of growth in investments in manufacturing was slower than in the case of other industries during the 1970s and early to mid 1980s, the pace accelerated during the late 1980s and early 1990s, outstripping the growth of FDI elsewhere in the U.S. economy (Borghese 1993; Rutter, 1991). Although foreign investments in manufacturing has leveled off somewhat in the mid-1990s, its relative share of total FDIUS remains higher than levels reached in the early 1980s (Tables I.3 and I.4).

Interestingly, most projections made in the mid-1980s about trends in FDIUS were wide of the mark. The most dramatic and unanticipated reversal of trend was in U.S. real estate. The crash in U.S. real estate prices during the late 1980s took the bloom off that rose by depreciating old investments and discouraging new ones. FDIUS in financial industries, including banking, surged in the mid-1990s, but earlier growth was certainly retarded by financial instability and banking crises in the United States and elsewhere. Finally, the growth in FDIUS in the areas of wholesale and retail trade has seemingly been steady throughout the 1980s and 1990s (Tables I.3 and I.4). However, even here appearances are deceiving because wholesale trade figures tend to be inflated. Based on the inappropriate methodology used to allocate industry statistics, the manufacturing of motor vehicles by foreign companies in the United States is actually included in the "wholesale trade" category rather than the "manufacturing" category (Rutter 1991, p. 26).

In short, the United States did become more of a service economy and less of a manufacturing economy during the time period under study. Interestingly, such a transition is not that visible in inward FDI statistics. Foreign MNCs continue to be attracted by opportunities to invest in the U.S. manufacturing sector.

The Quantitative Dimensions: The Relative Importance of Inward FDIUS

The data cited in the previous tables clearly reveal the sharp rise in inward FDI that has occurred since the early 1980s. However, how significant has this surge been in relation to other meaningful measures?

Table I.5 lends support to the notion that, for the United States, FDI was essentially a one-way street until the late 1970s. Although both the inward and outward FDI stocks grew during the 1950s, 1960s and early to mid-1970s, the former expanded from a much lower base. During this early period, the ratio of inward to outward FDI averaged only about 20 percent.

The extent to which inward FDI truly exploded in the 1980s is revealed by a dramatic rise in this ratio during the 1980s. In 1978, the United States' inward FDI position was only 25.1 percent of the outward position, but it rose to well over 90 percent by the end of the 1980s. Although the inward/outward FDI ratio

Table I.5. Foreign Direct Investment in the United States: in Relation to Capital Outflow Position and U.S. GDP

	FDIUS Inflows (Millions of US$)	*FDI Stocks (inward) (Million of US$)*	*FDI Stocks (outward) (Million of US$)*	*US GDP (Billions of US$)*	*Capital Inflows as % of GDP*	*Inward Position as % of Outward Position*	*Inward\ Position as % of US GDP*
1950	80	3,391	11,788	284.6	–	28.7	1.2
1954	124	3,981	17,748	363.1	–	22.4	1.1
1958	98	4,940	27,255	441.7	–	18.1	1.1
1962	132	7,612	37,145	556.2	–	20.5	1.4
1966	86	9,054	51,792	743,3	–	17.5	1.2
1970	1,030	13,270	75,456	974.1	0.1	17.6	1.4
1974	2,745	22,421	118.613	1,385.6	0.2	19.0	1.6
1978	7,874	40,831	162,727	2,107.0	0.4	25.1	1.9
1982	13,842	123,590	207,752	3,152.5	0.4	59.5	3.9
1986	34,091	220,414	259,800	4,230.8	0.8	84.8	6.5
1988	59,424	328,850	335,893	4,853.9	1.2	97.9	6.8
1990	37,213	403,735	426,958	5,464.8	0.7	94.6	7.4
1991	22,799	419,108	460,955	5,610.8	0.4	90.9	7.5
1992	18,885	427,566	502,063	5,920.2	0.3	85.2	7.2
1993	41,738	464,110	559,733	6,343.3	0.7	82.9	7.3
1994	50,060	504,401	612,109	6,738.4	0.7	82.4	7.5
1995	60,848	560,088	711,621	7,253.8	0.8	78.8	7.7

Source: *Survey of Current Business*, various issues.

has declined somewhat in the 1990s from the 1988 peak year level (97.9%), it has averaged approximately 85 percent, ranging from 78.8 percent in 1995 to 94.6 percent in 1990 (Table I.5). Without question, foreign direct investment in the recent U.S. experience has become a two-way street.

The growth in the relative importance of FDIUS is also revealed by comparing investment stock and annual capital inflow data to relevant macroeconomic yardsticks. As revealed by Table I.5, both FDI measures were tiny fractions of the U.S. GDP throughout the 1950s, 1960s and into the 1970s. The spurt in inward investment activity in the 1980s caused both ratios to rise. As recently as 1974, the ratio of FDIUS stock to U.S. GDP was only 1.6 percent but, because of the heavy infusion of foreign investment capital over the past two decades, the ratio has risen to consistently exceed 7 percent in recent years.

In reference to the ratio of annual capital inflows to U.S. GDP, the trend over the past two decades is also positive but less linear (Table I.5). Furthermore, FDI capital inflows have grown in recent decades in relation to domestic real investments in the United States. Following insignificant levels in the 1950s, 1960s, and 1970s, FDI capital inflows surged in the 1980s to the point that, by 1989, they represented 7.5 percent of domestic capital formation (OECD 1995, p. 11). Furthermore, in this same year, expenditures on new plant and equipment by foreign firms located in the United States accounted for 12.3 percent of total nonresidential gross private investment in the United States (Berzirganian 1991). Despite a slowdown in the pace of capital inflows and foreign investment expenditures in the 1990s, foreign capital continued to contribute significantly to the investment performance of the U.S. economy in the early and mid-1990s (OECD 1995).

Finally, the fact that the United States has become the leading host country for foreign direct investment has had important employment effects. For example, in 1989, a peak year for net foreign direct investment inflows into the United States, approximately 5 percent of the United States' workforce were employed by foreign held companies and nearly 10 percent of the manufacturing workforce owed their jobs to foreign held companies (Berzirganian 1991). In the 1990s, the conventional wisdom[7] persists that inward FDI have been, are and will continue to be important sources of job opportunities for American workers (OECD 1995).

The Quantitative Dimension: An Overview

Statistical evidence clearly reveals a quantum jump in FDIUS in recent decades. Although historically most accumulated FDI capital originated from Europe and Canada, Japan has become a major source of foreign investment since the 1980s, to such an extent that a special examination of Japanese motivations in this regard is warranted in the book.

The statistical examination of FDIUS by sector reveals that the phenomenon is multidimensional, not single dimensional. No one U.S. industry or industrial sector dominates as a magnet for FDIUS. Cyclical and structural change in the U.S. economy have affected the sectoral distribution of inward FDI in recent decades, but manufacturing, wholesale/retail trade, banking/finance, insurance and real estate all continue to attract both the attention of foreign direct investors and significant shares of FDI. Foreign MNCs continue to invest heavily in the U.S. manufacturing sector despite the "conventional" wisdom that the United States is in transition from a manufacturing to a service orientation.

Finally, the data, at least on the surface, suggests that FDIUS in recent decades has become increasingly more important in generating national income and as a source of new capital and jobs. Critics argue, of course, that aggregated data conceal as much as they reveal and that the social benefits of FDI tend to be exaggerated while the social costs are ignored. Given the complexities of underlying issues, public policymakers await clarification on the causes and consequences of FDIUS. It is the purpose of this book to shed light on the former.

NOTES

1. In a sense, Servan-Schreiber's prediction was conditional, that is, on the failure of Europe to meet the "American challenge." In his view, Europe was in danger of becoming an economic satellite of the United States only if European corporations continued to be uninnovative, if labor markets continued to be inflexible and if European government policy continued to produce market disincentives. Subsequent history would seem to indicate that the challenge was met.

2. Japanese FDI in the United States during the 1980s seemed to generate the most concern and alarm. See Burstein (1988), Prestowitz (1988) and Glickman and Woodward (1989).

3. For a comprehensive discussion of these issues, see Graham and Krugman (1995, pp. 57-84).

4. Although FDIUS was an issue in the U.S. Presidential campaigns of 1988 and 1992 as well as the more recent debate over NAFTA, arguments tended to be emotional not substantive and frequently appeared as "sound bytes" rather than empirically supported line of argumentation. See Ondrich and Wasylenko (1993, pp. 1-37).

5. Interestingly, in the United States, the individual states have been much more active and aggressive than the Federal government in offering investment incentives to potential foreign investors. Policies at both levels are examined in depth in Chapters V and VI of this study.

6. In reference to certain motivational areas, there is a blending of market and public policy factors that can not be separated. Despite the fact that some public policy issues do arise in Chapters III and IV, the central focus is on market determinants of FDIUS.

7. The conventional wisdom in this regard has been questioned by those who believe that FDI has little or no effect on aggregate employment in a country over the long term because of the conviction that employment is supply, not demand, determined.

Chapter II

The Development and Evolution of Foreign Direct Investment Theory

INTRODUCTION

Foreign direct investment (FDI) theory has been formulated over time in response to the challenge of explaining the nature, pace and direction of multinational firm investments overseas. Specifically, FDI theory has focused historically on three fundamental questions, namely:

1. Why do firms go overseas as direct investors?
2. How can foreign firms compete successfully with local firms, given the inherent advantage of local firms operating in the familiar, local business environment?
3. Why do firms opt to enter foreign markets as producers rather than as exporters or licensors?

In addition, as theory evolved, efforts have been made to explain why "big business" undertakes virtually all foreign direct investment and why foreign direct investment is much more important in some industries than in others.

The various streams in the evolution of FDI theory over the past four decades are identified in Figure II.1. The major contributors to the development of theory over time are cited as well.

As indicated in Figure II.1, microeconomics-based theories of FDI have evolved in two directions. Both fall under the umbrella of "industrial organization" theory. The dominant stream focuses on the internal attributes or characteristics of multinational cor-

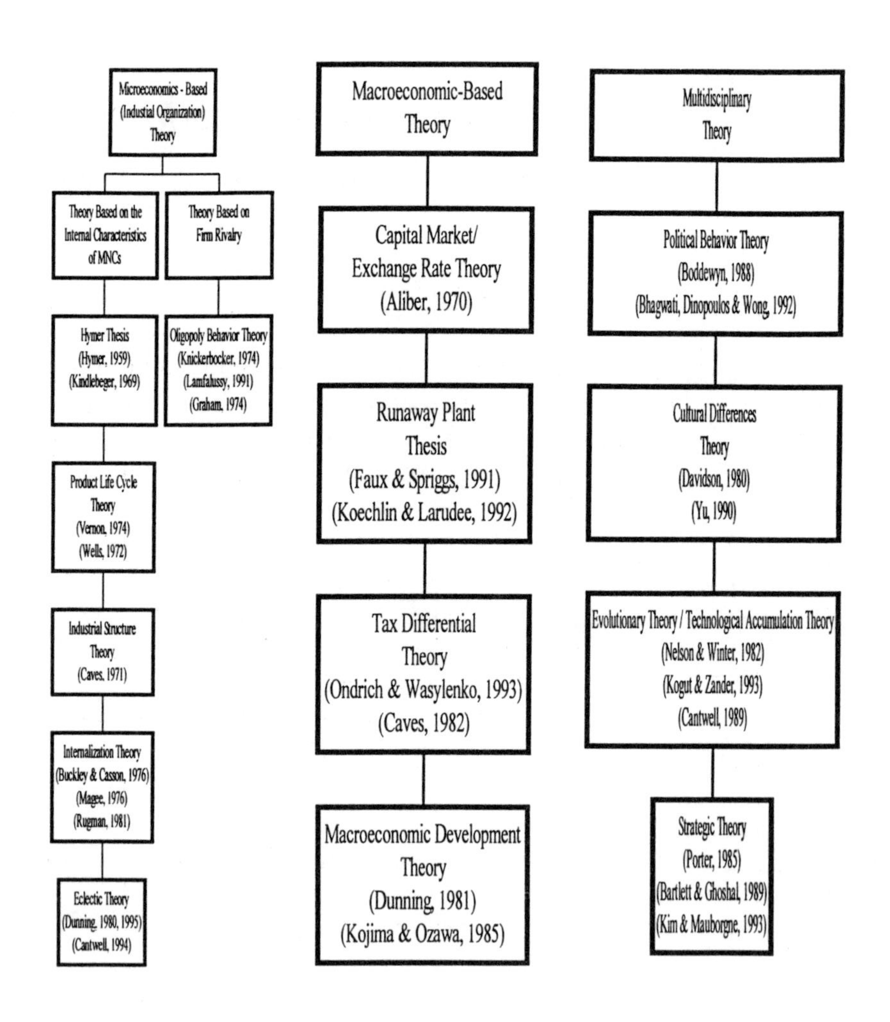

Figure II.1. Historical Evolution of Foreign Direct Investment Theory

porations that translate into competitive advantages overseas, while the second focuses more on intra-industry rivalry among firms and oligopolistic "follow-the-leader" behavior. An example of the former would be a firm that is willing to incur the costs of entering a foreign market as a producer because of a competitive advantage based on a superior product or process technology, and an example of the second would be a firm following an oligopolistic rival overseas to prevent an unfavorable swing in market shares.

A separate body of theory has evolved over the past four decades in the attempt to identify the macroeconomic determinants of FDI. The theoretical work here focuses on how differences in macroeconomic conditions or developmental stages from country to country may govern the pace and direction of FDI. Attempts at this level of analysis have been made to structure macro models capable of explaining FDI flows.

Finally, over the past two decades, new theories have emerged that are multidisciplinary in nature and which have been spurred by the recognition that economic factors offer only partial explanation of FDI. In this stream of research, two significant innovations are observable. First, the scope of FDI theory has broadened beyond the purview of economists to include the insights of scholars from the other social sciences as well as from the functional disciplines of management. Accordingly, attempts have been made to identify the political, cultural, social and managerial factors that govern FDI decision making as well as the economic factors. An example here would be a firm motivated to expand production activities overseas, not to exploit the advantage of a superior product or production process, but rather the advantages arising from political contacts or cultural ties in the host country.

The second significant innovation, observable in recent FDI theory, is the attempt to uncover the strategic factors that influence the growth and direction of MNC operations internationally. In the literature, there is a visible transition from earlier "static" analysis of why MNCs enter foreign markets as producers to the dynamics of how MNCs strategize to exploit their competitive advantage globally.

The several streams of FDI theory, identified in Figure II.1, will now be examined in detail with appropriate linkages noted.

MICROECONOMICS-BASED FDI THEORY: INTERNAL FIRM CHARACTERISTICS

Foreign direct investment theories developed in the 1960s and 1970s as logical extensions of microeconomic analysis of product and factor markets. It was generally held that FDI takes place because of market imperfections.[1] The imperfections in market organization or in the market adjustment process are the cornerstones upon which microeconomics-based FDI theories are structured.

Foreign Direct Investment and the Theory Of Monopolistic Competition: The Hymer Thesis

Early foreign direct investment theories borrow heavily from traditional theories of the firm that suggest, given the time/distance problems and the risks associated with foreign direct investment, international capital movements will not occur under conditions of perfect competition. It was Hymer's view (1976), as refined by Kindleberger (1969), that foreign direct investment belongs to the theory of monopolistic competition. Since the local firm has natural cost advantages based on location, the multinational firm must enjoy offsetting advantages based either on its differentiated product or gained through the capturing of scale economies.

Two conditions must exist to justify foreign direct investment. Imperfections in product and/or resource markets must permit the multinational firm: (1) to earn higher profits abroad than is possible at home and (2) to earn higher profits than local firms in host countries. This second condition would arise, of course, if the size of the local market in the host country were too small to allow local firms to capitalize on the scale economies captured by the multinational firm in extending its operations to the foreign country. In this context, of course, scale economies would arise from distribution or marketing advantages not production advantages. In the case of foreign direct investment in a large market area such as the United States, this investment motive would seemingly not exist. Local firms with access to the same large market would be in a position to capture the same scale economies as the foreign firm. Thus, if the Hymer thesis applies to this type of investment, the foreign firm's advantage would have to be based on product differentiation.

Product Life Cycle Theory

In the mid-1960s, Vernon added a dynamic dimension to the Hymer thesis, revealing how MNCs can capture, exploit and manage international comparative advantage over time. The product life cycle theory of Vernon (1966-1974) is an attempt to link comparative advantage in foreign direct investment to product differentiation.[2] In Vernon's view, this "differentiation" is based on superior proprietary knowledge which translates into a technologically superior product. Initially, product innovation by a particular company produces a differentiated product and a trade advantage locally. With the acceptance of the product in the domestic marketplace, foreign demand will arise at some point, and this will be satisfied initially through exportation. At this early product life cycle stage, the level of demand would not be sufficient to justify the establishment of foreign subsidiaries to manufacture the product abroad; however, foreign direct investment is one of three possible scenarios which may arise as foreign demand for the product grows.

The first scenario would be competition from local firms in the foreign country as market entry barriers diminish—ultimately causing the innovating firm to cede the foreign market to locals. Entry barriers would be originally based, of course, on the technology "gap" and resulting product differentiation. However, this gap would diminish with exportation, particularly if the product is capable of being reverse engineered.

Technology transfer would force the innovating firm in the home country to make a major decision if indeed it chose to continue to seek returns abroad on its investment in the original product concept. The decision might be to appropriate returns through a licensing arrangement with local firms (the second scenario) which would be consistent with that aspect of the product life cycle theory which predicts that the location of the most efficient plants would ultimately shift from the home to the foreign country.

On the other hand, the decision might be to compete by establishing manufacturing subsidiaries in the foreign country (the third scenario). This decision would also be consistent with that aspect of the product life cycle theory mentioned above. Two conditions seemingly must exist for the firm to select the direct investment alternative. First of all, the size of the foreign market must justify the investment outlay; second, there must be some remaining advantage based on product differentiation that is capable of offsetting the costs and risks of doing business overseas. If this early advantage erodes because of technology transfer, then the initial firm will predictably cede the foreign market to local firms. However, if some advantage remains but is perceived to be slight, the decision will probably be to license. Finally, if the advantage is perceived to be great despite the diffusion and dissemination of technology abroad, the logical business decision would be to support the foreign direct investment alternative.

Foreign Direct Investment and Industry Structure

Caves (1971) added an important new dimension to the theory of market imperfection by extending the Hymer thesis to incorporate the theory of industrial organization. Caves, in essence, argued that firms in consolidated industries tend to become multinationals because they obtain intangible assets from their investments in advertising and R&D.[3] Such firms create their own firm-specific, comparative advantage by producing and marketing differentiated products worldwide. Noting that multinationals typically come from industries characterized by both research and marketing intensity, Caves suggests that extending operations overseas is thus consistent with the desire on the part of a firm to maximize returns on heavy research and marketing expenditures.

In extending the Hymer thesis in this direction, Caves borrowed from the earlier work of Kindleberger (1969), who was the first to argue that MNCs may enjoy company-specific advantages in areas other than technology, such as managerial, organizational or marketing skills. As an example, a firm may decide to enter a foreign market as a producer, despite its failure to differentiate its product qualitatively through technological breakthroughs, if it is able to exploit advantages on the revenue side through the exercise of marketing skills or on the cost side through the capturing of organizational efficiencies.

Although market imperfections continued to serve as the cornerstone of foreign direct investment theory, a significant shift took place in the 1970s in the theoretical literature. In effect, the internal organizational characteristics of investing firms became the focus of attention. The historical antecedents of these new theories of internal organization were Coase (1937) and Williamson (1964).

Internalization Theory

A number of economists in the 1970s and early 1980s, most notably Rugman (1981) and Buckley and Casson (1976), questioned the Hymer conclusion that imperfect markets alone are sufficient to explain why direct foreign investment takes place.[4] After all, an advantaged firm could always exploit its superiority overseas by exporting or licensing activities. According to internalization theory, there must be economies arising from the exploitation of market opportunities through the firm's internal operations rather than through exporting or licensing for direct foreign investment to occur. This competitive advantage must be firm-specific, not easily duplicated and in a such a form as to permit that advantage to be transferred to foreign subsidiaries.

The key factor for maintaining such a firm-specific, competitive advantage is possession of proprietary information and control of the human capital that can generate new information through expertise in research, management, marketing and technology. According to this theory, information is an intermediate product par excellence. It is the oil that lubricates the engine of the multinational company. The expansion of the firm's operations overseas permits it to transform an intangible asset, namely information, into a valuable property specific to the firm (Rugman 1980, pp. 368-369).

In the view of Magee (1976, 1977), multinational corporations are specialists in the production of technical information. According to appropriability theory, which is a subset of internalization theory, this sophisticated information can more efficiently be transferred internally within organizations rather than through external markets. Borrowing from both the industrial organization approach to foreign direct investment theory and neoclassical notions about appropriating returns from corporate investments in information, Magee argues that firms opt to exploit market opportunities as direct investors because it is the best way to minimize transaction costs in the production of information and to appropriate maximum returns on its investments in new technologies. An example might be a high tech firm with a high ratio of R&D cost to total cost that is able to underprice competition globally if it is able to transfer information efficiently from parent company to overseas subsidiaries, thereby avoiding the higher costs and risks of information transfers through external markets.

According to Magee, capturing maximum returns on investments in information is particularly important for young industries because information is produced at a fast pace at early stages of the technological cycle. This leads Magee to argue, therefore, that younger, more innovative firms are motivated to serve external markets as direct investors, while more mature firms, generating information at a slower pace, see licensing as a more viable option.

With the maturing of direct foreign investment theory in the 1970s and 1980s, more attention was focused on the role of technology as the source of the multinational firm's comparative advantage as an overseas producer. Advances in product technology were viewed as the key in generating the superior "differentiated" product. Thurow (1992) contends, however, that in the future sustainable competitive advantage will depend less on new product technologies and more on new process technologies. According to Thurow, the major "pull" factor attracting multinational firms will be the education and skill of the work force. The ability of the foreign firm to exploit that skill will produce the competitive advantage.

Eclectic Theory

The eclectic theory of FDI incorporates the concept of internalization as part of three factors necessary to explain why MNCs opt to engage in international production as direct investors. Those factors, according to Dunning (1980, 1981b, 1988b), are: (1) the extent to which the firm possesses or controls assets that existing or potential rivals do not possess, (2) the extent to which it would be more profitable to internalize those assets in the firm's own production as opposed to selling or leasing them and (3) the extent to which it would be more profitable to exploit those assets internationally rather than domestically.

As a direct investor, the foreign firm must cover entry costs as well as the costs of serving an unfamiliar and distant environment. To succeed, the firm must possess "ownership" advantages sufficient to outweigh these costs. Such ownership-specific inputs may take the form of a legal right such as a patent, some form of market power such as input control (e.g., ALCOA cornering the bauxite market earlier in this century) or they may be based on superior, technological proficiency. The greater the degree of ownership-specific advantage possessed by a firm, the greater the inducement to internalize the advantage. If the foreign production base is more attractive than the domestic base, the internalization of ownership advantage will become foreign direct investment.

The eclectic foreign direct investment theory of Dunning maintains that, once a firm possesses net ownership advantages, it exploits these advantages in conjunction with the specific factor inputs existing in the foreign country. These location-specific factors, such as low materials or labor costs, in combination with the ownership advantage factors, explain why a firm will serve a particular market as a direct foreign investor rather than as an exporter or as a licensor.

In the 1990s, the eclectic paradigm underwent a period of reappraisal spurred by events that strongly suggested that the world economy had entered a new phase of market-based capitalism (Dunning 1995; Cantwell 1994). In this so-called age of Alliance Capitalism, trans-border activities became increasingly affected by the collaborative productive arrangements between and among organizational units of MNCs and other firms. Whereas apologists for traditional capitalism view collaboration as a symptom of structural market failure, those who embrace the virtues of Alliance Capitalism see it as a means of reducing endemic market failure (Dunning 1995, p. 466). Firm behavior has been affected by evolving political views on the acceptability of inter-firm alliances. Accordingly, the eclectic paradigm is in the process of being re-examined with the intention of incorporating into the model the role of cross-border alliances and inter-firm collaborative production and transactional arrangements in the new global economy.

MICROECONOMICS-BASED FDI THEORY: FIRM RIVALRY

All of the theories cited above share one common feature, namely that the source of comparative advantage that induces the MNC to enter a foreign market as a producer derives from some internal characteristic or quality of the firm. In the stream of research cited below, firm rivalry and defensive reactions are at the core of FDI theory. Whereas Hymer, Kindleberger, Vernon et al applied the theory of monopolistic competition in explaining foreign direct investment flows, the focus here is on oligopolistic behavior.

The observation that large U.S. based multinationals tend to follow one another into major foreign markets spurred the formulation of a direct investment theory by Knickerbocker (1974). According to Knickerbocker, patterns of foreign direct investment are consistent with traditional oligopoly behavior. A firm producing in an oligopolistic industry is compelled to follow a rival overseas even though the firm's assessment of the profit potential of the venture is far less optimistic than that of the rival. In effect, this oligopolistic follow-the-leader behavior is motivated by a desire to maintain market share stability within the industry by avoiding the possibility of any one firm dominating a new and potentially profitable market area.

This "defensive investment" strategy relates to the broader concept of Lamfalussy (1961), according to which firms investing overseas are motivated not by the desire to exploit a profitable situation in the short term, but rather by the desire to avoid getting shut out of a new market by aggressive rivals. Such firms, in the view of Lamfalussy, believe that it is preferable to enter a market with a low expected profit than to be excluded from it altogether. Aliber (1970) has also taken the position that foreign direct investment is a defensive measure taken by oligopolies to maintain foreign market shares in an increasingly competitive environment.

A related but different follow-the-leader effect was identified by Graham (1974), who observed an interesting investment pattern in which European investors in the United States seemed to imitate the activities of U.S. investors in Europe. Increased U.S. direct investments in Europe appeared to spur a reverse flow of investment capital after a relatively short lag.

In reference to investment motivation, he hypothesized that:

> ...such strategy of cross-investment by European firms would undermine the possibility of U.S. firms engaging in a strategy of price cutting in the European market to gain market share while maintaining price stability in the United States. If European firms were to enter the U.S. market, they would achieve an "exchange-of-threat." Aggressive tactics in one market could be met with counter-aggression in the other market (Tsurumi 1977, p. 84).

According to Graham, there are strong feelings of mutual interdependence in direct investment decision making that extend internationally and across industries, traditionally defined.

MACROECONOMICS-BASED FDI THEORY

One common thread extends through the theories developed in this stream of research. They all focus on differences in macroeconomic conditions or developmental stages among nations in explaining what motivates MNCs to extend their productive operations overseas. However, these macroeconomics-based theories do have microeconomic roots and do borrow from the theoretical work examined earlier. Both microeconomic and macroeconomic models are crafted from the same cornerstone, namely, the assumption that foreign direct investment takes place ultimately because of market imperfections. Such "imperfections" are examined at different levels of analysis. On one hand, microeconomic theory focuses on the internal characteristics of firms operating globally under conditions of imperfect competition or on intra-firm rivalry in consolidated industries. On the other hand, macroeconomic theory examines how market imperfections, particularly in capital and labor markets, create opportunities for MNCs to exploit differences in macroeconomic conditions overseas.

Foreign Direct Investment and the International Capital Market

Several macroeconomics-based theories of foreign direct investment focus on capital market behavior. Aliber (1970, 1983), argues that direct investment flows are governed by exchange rate risk and capital market imperfections. Specifically, he suggests that direct investors in stronger currency areas will have a comparative cost advantage over investors in weak currency areas if lenders (liquid asset holders) are willing to pay a premium in order to hold assets denominated in the stronger currency. The resulting premium is reflected in an effective interest rate differential that is equal to the premium plus the expected rate of appreciation of the stronger currency.

Given this interest rate differential, the investor from the strong currency area is able to secure capital at a lower cost and to weigh investment opportunities at a lower discount rate than possible for the weak currency investor. Given capital market imperfections, investment opportunities in weaker currency areas will be more attractive to strong currency investors than to weak currency investors if the net present value of income, generated in the weak currency area is greater.

A second example of a foreign direct investment theory based on capital market behavior is the diversification of portfolios model of Levy and Sarrat (1970). According to this model, firms become "multinational" by making direct investments overseas in order to offer to their stockholders the advantages of holding claims on both domestic and foreign income streams. Theoretically such a "portfolio" would contain more optimal risk/return characteristics than one derived from a domestic income stream only. If this result holds true, multinational firms would enjoy a comparative advantage over local firms, based on a lower cost of capital, and would therefore be in a position to evaluate income streams at lower discount rates.

The issue of international diversification as a risk reduction measure is, of course, of interest to all investors. The multinational firm is no exception in this regard. Lessard (1979) postulates that the multinational firm has a unique advantage in international diversification through its ability to internalize financial transactions. Specific gains to the firm accrue from exchange control arbitrage, credit market arbitrage and equity market arbitrage.

In the 1970s and early 1980s, direct investment theories based on capital market imperfections were hardly in the forefront of the literature. This neglect was based on the assumption that financial markets tend to be more efficient than markets for real goods and services. Recent evidence of financial market imperfection has produced some renewed interest in the literature. For example, in a 1990 study, Madura and Whyte argue that multinational firms capture significant risk reduction benefits through international diversification. Strategies are designed to stabilize cash flows, thereby reducing the risks perceived by shareholders and creditors. This, of course, translates into favorable effects on capital costs. Foreign direct investments allow firms to diversify product lines in order to insulate cash flows from industry-specific events.

Recently, there has also been renewed interest in the impact of foreign exchange rate movements on FDI decision making. As indicated above, earlier studies focused on the level of exchange rates, arguing that foreign investors from strong currency areas are able to secure capital in host countries at relatively low costs. The linkage between real exchange rates and relative production costs globally continues to be a focal point of more recent research, but other areas of interest have emerged in the literature including: (1) the impact of fluctuations in real exchange rates on FDI through the altering of relative wealth across countries (Hultman and McGee 1988; Klein and Rosengren 1994) and (2) the impact of exchange rate volatility on FDI decision making (Campa 1993). In reference to the latter, it is argued that the direction of exchange rate movement, as a factor governing FDI, is less important than the degree of volatility in its fluctuation. High volatility promotes uncertainty and risk, thereby discouraging the movement of capital across international boundaries.

Foreign Direct Investment and the Runaway Plant

An alternative view is that FDI is driven more by labor market, than by capital market, imperfections. The term, "runaway plant," was coined in the 1970s to describe the manufacturing subsidiary of the multinational company, established overseas for the purpose of minimizing labor costs. Foreign direct investment in this setting is motivated by cheap foreign labor. United States proponents of the anti-multinational Burke-Hartke bill in the 1970s used this concept to argue for artificial barriers to block the so-called "exportation of jobs." The same argument echoed in the early 1990s as articulated by opponents of NAFTA.

Clearly, the theoretical basis of this position rests on the assumed existence of labor market imperfections. If the market were perfect, relatively low wages abroad, reflecting low labor productivity, would not produce lower labor costs per unit of output. On the other hand, assuming labor market imperfections, substantial differences in the wage rate/labor productivity ratio may exist. If the ratio is significantly lower in the foreign country, labor cost advantages of investing abroad may be sufficient to offset high non-wage costs, thereby providing the multinational corporation with the necessary incentive to invest in so-called "runaway plants."

Opponents of NAFTA such as Koechlin and Larudee (1992) and Faux and Spriggs (1991) in the early 1990s used the imperfect labor market argument in forecasting a heavy flow of FDI capital from the United States to Mexico following the ratification of the treaty. It was also suggested that the consequences would be adverse for the United States. This specific complaint was part of a larger argument used by those who fear that low-wage LDCs in a free, "unfettered" international capital market would be able to attract both capital and technology from industrialized countries, achieving higher levels of productivity, while paying appreciably lower wages. It is argued that industrialized countries would be victimized in the process with balance-of-payment deficits and with either large-scale unemployment or sharply declining wages.

Are MNC runaway plants motivated to minimize production costs by seeking out investment only or mostly in low wage nations? FDI data presented in this book and elsewhere do not support that contention. Furthermore, the concept of the runaway plant and the scenario above seem to violate some basic economic principle as pointed out by Krugman (1996, Chap. 5) and Hufbauer and Schott (1993, Chap. 2). Certainly, at some point in time, wages in relation to labor productivity may be lower in an LDC than in the case of an industrialized nation, but this difference will not persist over time. As capital and technology flow to low-wage countries, their wage rates will rise along with their productivity. It is naïve to assume that capital importing nations may be blessed with productivity advances while maintaining their low-wage status. Also, it would be a mistake to assume that capital importing LDCs would be able to maintain trade surpluses with capital-exporting industrialized countries. Capital imports permit a country to invest domestically in excess of domestic savings, and this will swing balance of payments into deficit disequilibrium (Krugman 1996, p. 76). Industrialized countries will not be victimized over time by becoming net exporters of FDI capital.

Can the "runaway plant" thesis explain why FDI tends to flow in a particular direction at a particular time? The answer is "yes," but it is unlikely that the motivating condition, that is, the wage rate/productivity differential between the host and investing countries, will persist over time given market adjustments. Furthermore, labor market imperfections may explain why an MNC might be motivated to invest in a certain host country at some point in time rather than invest at home, but they do not explain why the MNC would have a comparative advantage over local firms in this regard.

Foreign Direct Investment and National Tax Differentials

Macroeconomic-based theories, such as those examined above in the previous two sections, argue that FDI is governed by capital and labor market imperfections that create production cost advantages for firms operating overseas. It may be argued, of course, that it is the bottom line, net of taxes, that is the "carrot at the end of the stick," enticing the firm to exploit overseas investment opportunities. Relatively low production costs may promote profitability but the advantage here for the foreign firm may be offset by unfavorable host country tax treatment. In reference to national tax policy, both economists and business leaders generally embrace the concept of a "level playing field," but the term typically has different implications for the two groups.

Economists are more inclined to focus broadly on the impact tax policy has on economic welfare and on efficiency in the global allocation of capital, where foreign businesses are more interested in the effect of tax policies on their net profits compared to those of their local competitors. Economists fear tax policy that misallocates capital globally where business people fear tax policy that unfairly favors local competitors. Interestingly, it is the same fear, articulated from a different point of view. World economic welfare would be maximized if tax systems are neutral in the sense of having no impact on the locational decision making of MNCs globally (Ondrich and Wasylenko 1993, pp. 15-16).

One stream of economic theory does explore ways to maximize world output and return to capital through a completely neutral tax system for capital, referred to in the literature as "capital-export neutral" (Caves 1982; McLure 1992). However, while economists structure tax models in a perfect world setting from an economic welfare point of view, businesses are inclined to favor territorial tax systems known as "capital-import neutral," under which foreign MNCs pay taxes only to host country governments and not to home country governments.

Unfortunately, MNCs operate in a global economy which is neither perfect from the economist's nor the business leader's point of view. Tax differentials exist among nations and among political subdivisions within nations, and most tax systems are neither capital-export nor capital-import neutral. Although no one would argue that tax differentials play the dominant role in FDI decision making, particularly in light of the importance of industrial organization considerations in this regard, once the decision has been made to invest overseas, taxation can have an impact on the specific location of the investment (Ondrich and Wasylenko 1993, p. 14).

In an imperfect world, complexities exist because different host countries impose different types of taxes on foreign investments, and in some host countries, for example, the United States, layers of taxation are imposed on foreign investment and foreign source income by political subdivisions, such as states. Furthermore, some home countries employ a territorial taxation system, not taxing the foreign source income of its MNCs, while others employ a worldwide taxation system, taxing the income of overseas subsidiaries, but granting credits for taxes paid to host country governments (Graham and Krugman 1995, p. 48).

Interestingly, worldwide taxation systems can produce an unexpected effect if host countries offer tax holidays or cut taxes on local income as incentives to attract FDI. Tax measures of this type would logically succeed if MNCs are from home countries that employ territorial taxation systems because foreign source income is not taxed and, thus, there is no possible loss of tax credits at home. On the other hand, with a worldwide taxation system, a host country tax cut may put MNCs at a competitive advantage vis a vis local firms because the former lose tax credits at home and the latter do not.

Accordingly, it may appear on the surface that host country tax cuts produce FDI incentives but, in reality, with a worldwide tax system, the effect may be neutral or may produce disincentives, depending on circumstances. There is a linkage between international tax policy and FDI decision making, but the effects are complex, not simple. From the literature, it is apparent that tax differentials among nations matter in influencing the flow and direction of FDI, but they are of secondary importance in comparison to industrial organizational considerations contained in microeconomics-based theory.

Macroeconomic Developmental Theory

As evidence that macroeconomic theories of FDI do have microeconomic roots, a body of theoretical work appeared in the literature during the 1980s as a logical extension of the product life cycle model of Vernon (1966, 1974) and the eclectic model of Dunning (1980, 1995). Where Vernon asserts that FDI is governed by the various development stages of a company's product life cycle, proponents of macroeconomic development theory argue that the nature, pace and direction of FDI are influenced more by the particular stages of economic development that the investing and target countries are passing through.

According to Kojima and Ozawa (1985), the rapid growth of industrialized countries spurs the outward flow of FDI. As local firms in advanced countries innovate, achieving higher levels of technological sophistication, they become increasingly more motivated to locate their less sophisticated types of production and technology in countries at earlier stages of economic development. The products and processes rendered obsolete by rapid technological progress at home still have market value in lesser developed countries. Clearly, this is a macroeconomic version of a product's life cycle. Maturing lines of production, losing value at home, are relocated to countries which are a step or two behind in the developmental process.

Borrowing from the Japanese experience, it is also argued that countries experiencing rapid growth also may be constrained by local, natural resource availability. Accordingly, such countries would be motivated to invest in resource related ventures overseas in order to supplement their own domestic supplies. Furthermore, with investments in resource extraction industries in lesser developed countries, MNCs from industrialized countries would also have a direct interest in moving basic resource processing and related manufacturing activities close to extraction sites (Cantwell 1989, pp. 201-202).

In drawing macroeconomic implications from his own eclectic theory, Dunning (1981b) argues that inward FDI, outward FDI and the balance between the two are all linked to a nation's economic development stage. In fact, Dunning identifies a four-stage FDI developmental cycle. In stage one, little or no inward or outward FDI take place in a developing country because the local investment climate is unattractive and local companies lack comparative advantage globally. However, inward investments arise in stage two with improvements in infrastructure and social overhead capital. With maturing industrialization, outward FDI begins in stage three as local industries gain some competitive advantages in world markets, but net investment flows remain inward. Finally, the country becomes a net capital exporter, that is, outward FDI grows to exceed inward FDI, as it reaches the fourth and final stage of development (Dunning 1981b, p. 136).

The linkages between eclectic theory and macroeconomic development theory are clear, and together they explain why countries at advanced stages of development tend to be net capital (FDI) exporters. The ability to create and sustain technological and human resource superiority increases as countries reach higher levels of development and this process bestows ownership-specific advantages on local firms as they enter into global competition.

MULTIDISCIPLINARY THEORY

All of the theories examined above have one common characteristic, that is, the central core of each explanation of why FDI takes place falls well within the normal boundaries of economic analysis. Since the 1980s, however, FDI theories have emerged which are more multidisciplinary in nature focusing on the political, cultural, social and managerial factors governing FDI in addition to the economic. At the same time, theories began to focus less on the static issue of entry and more on the strategic factors promoting MNC growth throughout the global economy. These theories are examined below.

Direct Foreign Investment Theory and Political Behavior

It should be emphasized that multidisciplinary theories did not develop in isolation. Economic theory, both on the macroeconomic and microeconomic side, has provided the foundation upon which these relatively recent paradigms are based. Figure II.1 indicates the paths down which FDI theory has evolved vertically. There has been a horizontal transfer of knowledge as well with economic theory serving as a springboard. For example, in Dunning's eclectic theory, the notion of ownership advantage is linked to the notions of "knowledge" or "expertise." This knowledge is essentially of the "economic" type, including both technical knowledge and management skill.

In the 1980s, ownership advantage in the literature was extended to include "political" knowledge and expertise.

> Such political advantages can take the forms of: (1) better intelligence about political actors and opportunities; (2) readier access to political opinion and decision makers, and (3) superior influence skills at handling the latter through various means. These political resources are "intermediate products" whose products may be internalized and exploited by the MNE in foreign locations (Boddewyn 1988, p. 345).

In this body of theory, there has been some focus on political risk analysis and on the link between political stability (or instability) and foreign direct investment (or disinvestment). There have also been analyses of (1) the interaction between government policies and global strategies, (2) the organization of external affairs in multinational companies and (3) negotiations between governments and multinational companies. These complex relationships are explored by use of political economy paradigms which examine the roles of all those social actions (including government) in the investment process that possess power, control or influence over relevant economic resources.

One interesting aspect of the government policy/global business strategy interaction involves the influence of the commercial policies of national governments, free trade associations and regional common markets on foreign direct investment flows. Tariffs and other barriers to trade restrict the least-cost supply of goods to foreign markets. Manufacturing firms, faced with these barriers, are motivated to enter foreign countries as producers rather than as exporters (Stevens 1973; Horst 1971). A lowering of trade restrictions, of course, is predicted to have the opposite effect.

Early proponents of this theory were quick to point out that the imposition of E.U. trade barriers against U.S. exports during the early and mid-1960s seemed to precipitate an accelerated and sustained increase in direct investments in Europe, particularly in the manufacturing area. Speculation continues that, if European integration in the 1990s under the Maastrict Treaty does produce a Fortress Europe effect on trade, another surge of U.S. foreign direct investment will be forthcoming.

Where the restriction of commodity trade may serve as a stimulant to foreign direct investment flows, the imposition of certain resource barriers may serve as a deterrent. Of signal importance in this regard would be capital barriers imposed by the government of either the host country or the investing country. Such barriers would block entry (or exit) in the first instance, or if imposed on new capital flows in support of established investments, could prompt disinvestment by reducing the degree of freedom in the financial management of multinational corporations. Similarly, restrictions on earnings repatriations could serve as a disincentive for new investments and could prompt disinvestment. Finally, if the overseas subsidiary of the multinational company is dependent on the importation of manpower from the home country, work permit restrictions and similar constraints may also serve as a disincentive.

Commodity trade restrictions, particularly those applied to finished goods, serve to stimulate foreign direct investment. However, the import restriction of raw material or intermediate goods by the host country may be a deterrent if a subsidiary of a multinational company, operating in that country, is dependent on such imports from the parent or seeks to benefit from the parent's monopsonistic position in the markets for such inputs.

Recently, foreign direct investment theory has evolved in the exploration of the multifaceted political linkages between multinational firm activities and the economic policies of the governments of host countries. Rather than merely weighing the impact of government policy (e.g., tariff policy) on direct foreign investment flows, FDI theory has expanded to measure the dynamics of the relationship. As argued by Bhagwati, Dinopoulous and Wong (1992, p. 186), foreign direct investment theory required a refocusing "so as to include explicitly the fact that the policy framework of the host country can be endogenous to the direct foreign investment in ways that needed to be formerly incorporated in the positive and normative analysis of FDI."

Since multinational firms gain ownership and control over that in which they invest, to some extent they limit the ability of host governments to chart independent domestic policies. However, multinationals transfer an array of assets (management skill, capital, technical know-how, etc.) to the host country. Thus, the challenge to the host government is to tap these assets without significant loss of political independence.

One technique examined in the literature by Vaitsos (1974) and Streeten (1974) involves efforts by host country governments to "unbundle" direct investment packages by purchasing elements of the investments through contracts, for example, capital through bank loans, technology through licensing, marketing services through distribution contracts, and so forth. A second technique, chronicled in the literature by Oman (1984) and Hennart (1989) involves new forms of investment. "New forms . . . include arrangements that fall short of majority ownership, such as various forms of contracts [licensing, franchising, management contracts, turnkey and product-in-hand contracts, production-sharing contracts and international subcontracting] as well as joint ventures" (Hennart 1989, p. 212).

In negotiating these arrangements, of course, governments seek to extract larger shares of the rents that foreign direct investments generate. According to theory, the degree of flexibility or rigidity by which host governments negotiate determines in part the direction and locations of the ultimate investments.

Finally, Thurow (1992) has added a new dimension to the political theory of foreign direct investment by arguing that corporations today offer the best opportunities to nation states to engage in empire building. Citing the Japanese as a case in point, Thurow asserts that the multinationals of some countries are motivated to go overseas, not as profit seekers, but rather to implement national strategies and to promote the self-interest of the nation state. Thus, investment motivations are political as well as economic.

Foreign Direct Investment and Cultural Differences

Efforts were made in the 1970s and 1980s to identify the expansion paths that firms follow as they move their production facilities overseas. Cultural similarities and differences were identified as factors that govern, in part, the location decisions of multinationals. In a 1980 article, Davidson (p. 18) hypothesizes that "firms in the initial stage of foreign expansion can be expected to exhibit a strong preference for near and similar cultures. Those in advanced stages of foreign operation will exhibit little, if any, preference for near and similar cultures." There is some empirical evidence in support of this hypothesis. For example, it is revealed in Chapter VI that European and Japanese multinationals do tend to cluster in certain regions of the United States. In part this is attributable to differences in state investment incentive programs. In larger part, it is attributable to the tendency on the part of relatively new FDI entrants to form cultural clusters.

The internationalization process is seen to accelerate as firms gain experience operating in different cultures. Psychological barriers gradually break down, but it can be a slow process, particularly for the inexperienced firm. Johanson and Wiedersheim-Paul (1975) and Johanson and Vahlne (1990) see the internationalization of the firm as a process consisting of a series of small steps, whereby firms gradually expand their operations as they gain operating experience in unfamiliar settings. Experience is gained through a process of "learning by doing." Yu (1990) argues that experience comes in two forms: (1) country-specific experience and (2) general international operations experience.[5] It is in the first area that country-by-country cultural differences govern the pace and location of the firm's internationalization.

Evolutionary Theory

As FDI theory became more multidisciplinary in the 1980s and 1990s, borrowing from organizational, social and political theory, as well as from economic theory, a new way of viewing the multinational corporation emerged. In this literature, called evolution theory, the growth and development of global firms was viewed as an evolutionary process beginning with market entry and extending through the exploitation of comparative advantage existing in the form of superior product knowledge.

In short, the multinational firm is viewed as a social community whose productive knowledge fully defines its comparative advantage. A firm's success in international profit seeking, therefore, is closely linked to the evolutionary acquisition and recombination of knowledge and the degree of efficiency by which knowledge is transferred globally through its network of subsidiaries and from parent to subsidiaries.

The pioneering work on the evolutionary theory of the firm is Nelson and Winter (1982). Building on this basic foundation, Kogut and Zander (1993) have extended these ideas within the context of corporate knowledge and know-how in examining the evolutionary growth of the multinational company globally.

In a sense, Kogut and Zander borrow from earlier formulated paradigms that either focus on the importance of knowledge in producing ownership advantages for multinational firms (Caves 1982; Magee 1976) or stress the strategic aspects of MNC planning (Porter 1986; Kim and Mauborgne 1993). However, evolutionary theory rejects the notion that multinational firms owe their existence and success to market failure. In rejecting the market failure approach, Kogut and Zander question the notion that firms exist to "internalize" markets. Knowledge and know-how do provide the key firm-specific ownership advantages that gives multinationals the competitive edge. Although knowledge transfer is important, it is decidedly of secondary importance, according to evolutionary theory, whether that transfer takes place internally or externally.

The pioneering work of Cantwell (1989) has produced a subset of evolutionary theory, referred to in the literature as the theory of technological accumulation. Cantwell argues that it is necessary to shift away from an "industrial organization" approach to more of an "industrial dynamics" approach in order to explain changes in the pattern and flow of FDI.

Decisions to invest in overseas markets are linked to the level and degree of technological competition among firms in global industries and this should be thought of as an evolutionary process. Firms in home industries with strong innovative traditions are motivated to invest overseas in countries which are themselves centers of strong innovative activity for two reasons. First, in such host countries there is much to learn from horizontal and vertical technological transfers, and the innovative firm is motivated to embark on a course of MNC expansion in order to capture these economies. Second, the investing firm's internal technological capabilities promote a competitive advantage overseas. Organizational research, development and engineering skills permit the firm to gear up to the competitive challenges of operating in a foreign market where technology is advancing rapidly (Cantwell 1989, pp. 2-3).

Technological accumulation theory is useful in explaining the logic behind the establishment of MNC production and research networks globally and in identifying the dynamics of competition among MNCs, particularly in technologically sophisticated, industrialized countries. According to theory, FDI decision making is governed by the technological histories of investing firms and, in the final analysis, the spread of international productive capabilities of MNCs is closely linked to global patterns of technological accumulation. FDI essentially is a by-product of MNC strategies, seeking essentially to exploit existing technological advantages and to seek new advantages.

The International Strategy of the Multinational Firm

Foreign direct investment theory matured in the 1980s, assuming more of a strategic and less of a static focus. Whereas early theories examined the processes by which firms become multinational, Porter (1985, 1986) and his Harvard Business School colleagues structured their new theory around the strategic issues for established multinationals. Porter distinguished multidomestic firms, in which a competitive presence in one country is independent of its presence in any other country, from globally or geocentrally-oriented multinationals that operate extensive networks of foreign affiliates with a high degree of centralized coordination. In the Porter model, multinational firm strategies are examined, based on the extent to which cross-border, value-adding investments are coordinated and on whether firm growth and expansion are centralized or decentralized.

If a multinational firm adopts a strategy based on the coordination and centralization of operations, globalization results. To Porter this means that the world becomes a strategic chess game to the firm's managers. Expansion of the firm's operations into one country is linked to its presence elsewhere; the firm's competitive position in one area is significantly influenced by its competitive position elsewhere. Organizational units do not seek to maximize profits in isolation. Rather, they are managed strategically with the objectives of the total organization foremost in mind. To Porter an industry is global, not multidomestic, if there is some competitive advantage to integrate activities on a worldwide base. Given that competitive advantage, firms' activities must be examined within the framework of a strategic model, focusing on the total, not the individual parts.

In the late 1980s and 1990s, interest in the strategic issues of multinational firm activity grew. There was a shift in emphasis away from the corporate "overview" position towards a focus on the multinational subsidiary as a unit of analysis (Birkinshaw and Morrison 1995). The various strategic roles that subsidiaries could assume became a focal point in international management research (Bartlett and Ghoshal 1989).

Recently, there has also been a shift in emphasis in FDI paradigms rooted in the principle of comparative advantage. Advantages based on differences in resource endowments are considered to be of secondary importance to those based on an organization's strategic capability. The multinational with the greatest ability to conceive and execute complex strategies is given the best chance to capture markets and profits in the global economy.

One method used in the successful development of competitive advantage has been through an examination of the dynamics of the strategy-making process between multinational firm head offices and subsidiary units. Early studies in this area focused on "what" the dimensions of an effective strategy might be, whereas more recent works have begun to raise the question of "how" global firms both conceive and execute strategic decisions (Kim and Mauborgne 1993).

Questions and issues of this type reflect how far FDI theory has gravitated in its evolutionary process from the 1950s and 1960s when economists debated within rather narrow parameters the motivational factors that induce firms to make the initial decision to move productive operations to overseas markets.

FOREIGN DIRECT INVESTMENT THEORIES: SUMMARY AND CONCLUSIONS

Foreign direct investment theory, although considerably younger than international trade theory, borrows from the same rich economic literature on competition and comparative advantage.[6] The development of FDI theory was spurred in the 1950s and 1960s by the challenge of explaining heavy U.S. foreign direct investments overseas during this period. Why did U.S. firms opt to go overseas as producers, and what was the source of their comparative advantage?

Early theories that developed as logical extensions of the theory of the firm had strong microeconomic roots. Given market imperfections, these theories focused on the investing firm's ability to exploit intangible assets, thereby offsetting entry costs and capturing comparative advantage in the process. Analysis was either static (Hymer, Caves, Kindleberger) or dynamic, as in the case of Vernon's product cycle model.

As industrial organization explanations of foreign direct investment theory matured, two main streams emerged: the first concentrating on the internal organizational qualities of investing firms and the second focusing on competition and rivalry among such firms.[7]

In the former stream, economists, such as Dunning, Buckley, Casson, Magee and Rugman attempted to pin down the firm-specific advantages that a multinational firm has by emphasizing how the firm internalizes transactions to capture comparative advantage, particularly in the market for knowledge.

Intra-industry rivalry is the focus of the second stream in the industrial organization literature. Borrowing more from the theory of oligopoly (as opposed to the theory of monopolistic competition), Knickerbocker, Flowers and Graham see foreign direct decision making as an action/reaction dynamic with clear cut "follow the leader" patterns.

As an alternative to the microeconomics-based approach, some economists historically opted to analyze factors governing the FDI decision making process more at the macroeconomic level, that is, by examining broad national and global trends and developments. To Aliber, Caves, Ondrich, Wasylenko and others, FDI flows are governed by differences in macroeconomic conditions and trends among nations in the global economy, created either by imperfections in resource markets or by macroeconomic policy initiatives. It should be emphasized that the microeconomic and macroeconomic approaches to FDI theory are not totally separate and distinct. Most macroeconomic theories have microeconomic roots and some paradigms, such as the Vernon product life cycle model, examines FDI decision making at both levels of analysis.

Developments in the 1970s and 1980s led to the evolution of foreign direct investment theories of a more multidisciplinary and multidimensional nature. As the field of international business matured, the dominance of economists in the formulation of theory diminished. In the literature, the multinational firm became the subject of study and research by scholars and practitioners in the functional disciplines of management, such as organizational theory, international finance and international marketing.

Also, the recognition grew that, although there are a wide variety of economic factors that both push and pull the foreign direct investor, there are political, cultural and social factors as well. Accordingly, the conviction emerged that understanding the multitude of economic and non-economic factors involved in the investment decision required a multi-disciplinary approach. Based on the premise that theories narrowly focused on "economic" factors alone were underspecified, attempts were made by Dunning, Boddewyn, Vaitsos, Streeten, Oman, Hennart, Davidson and Yu, among others, to elevate FDI theory to more of a "politico-economic" and "socio-cultural" level.

More recently, studies by Porter, Bartlett, Ghoshal, Kim and Mauborgne employ a strategic approach in the examination of the multinational firm. Whereas earlier FDI theory focuses more on the requisites of becoming a multinational firm, Porter et al. identify the strategies that established firms follow to continue to exploit their competitive advantage. Kogut, Zander and other proponents of evolutionary theory see the multinational firm as a social community specializing in the transfer and recombination of knowledge and strategizing to translate knowledge superiority into comparative advantage. In a related way, Cantwell sees linkages between FDI decision making and the exercise of technological virtuosity of innovative firms. FDI is viewed as part of an evolutionary process whereby investing firms are motivated to establish production and research networks overseas to exploit existing advantages and to accumulate new technologies through horizontal and vertical transfers. Technological and capital accumulation run alongside one another and the relationship is viewed as being symbiotic.

The multidisciplinary strategic approach to FDI theory has set the agenda for future studies. A consensus has clearly emerged in the literature that it is too limiting to view foreign direct investment as a decision made at some discrete point and to build theory around that decision making process. The decision to enter is only the first step in the process. The strategic dynamics of the entire process of operating a firm overseas must be incorporated into the theoretical construct. Recent studies on international management have taken an important first step in this regard by providing valuable insights into what the qualities of an effective worldwide strategy for multinationals should be. However, much more attention must be focused in the future on how global firms can actively both conceive and execute strategic decisions.

The question remains, of course, whether any or all of the theories reviewed in this chapter offer complete, or at least partial, explanations of why FDI takes place. Which theories, if any, are empirically verifiable and which are not? It is the purpose of the next two chapters to present the evidence and to pass judgment.

NOTES

1. For a comprehensive summary of the early FDI theories based on product and factor market imperfections, see Tsurumi (1977).
2. Also see Wells (1972) and Hirsch (1967).
3. For a refinement of this point, see Caves (1982).

4. The early formulation of internalization theory is explored in greater detail in Rugman (1986).

5. For a comprehensive review of this literature, see Bonito and Gripsrid (1992).

6. Interestingly, there have been early attempts to apply so-called "new trade theory" to the area of foreign direct investments. See Markusen (1995) and McCulloch (1993). In new trade theory, gains from trade arise independently of any discernible pattern of comparative advantage as multinational companies exploit scale economies and employ strategies of product differentiation in imperfectly competitive environments. The transition from trade to investment theory has been slow, however, since the literature on the former is very limited in the treatment of the firm as an entity much beyond that of a mere production facility.

7. A concise summary of both streams appears in Graham and Krugman (1995, pp. 191-193).

Chapter III

The Determinants of Foreign Direct Investment in the United States: Factors Governing the Investment Decision

INTRODUCTION

As indicated in Chapter I, one of the most dramatic developments in the global economy during the decades of the 1970s and 1980s was the sharp rise in foreign direct investment in the United States. Direct investment inflows during the first two post-World War II decades paled by comparison to the outflow sponsored by growing U.S. multinationals. During this period, foreign direct investment was, indeed, a one-way street. From the late 1960s to the late 1980s, however, the United States became increasingly more attractive as a host country, and the impact of capital infusions was significant both in terms of internal and balance-of-payments effects. Although capital inflows diminished somewhat in the early 1990s, the decreased propensity for investment in the United States was rooted in the cyclical weaknesses of the period and in random developments such as the collapse of U.S. real estate prices. This cannot be viewed as evidence of a reversal of trend, since the mid-1990s have witnessed a resurgence in FDIUS.

This chapter is concerned with the identification and examination of those motivational factors which spurred the growth of foreign direct investment in the United States since the 1960s.

This analysis is divided into two distinct time periods. Following a brief review of early FDI theory, the results of empirical studies, designed to identify the factors and/or conditions that induced foreign firms to invest in U.S. property assets during the 1960s and 1970s, are examined and analyzed. The same questions as the above are then raised in reference to the most recent infusion of foreign direct investment during the 1980s and 1990s.

The rationale for the two-period comparative analysis is that significant structural changes have been taking place in the U.S. economy and in the global economy since the 1980s, and accordingly, the factors that "pulled" and "pushed" investment capital from overseas during the earlier period may be somewhat different from the factors governing the investment decision more recently. It is the central purpose of the chapter to ascertain whether such differences exist and, if so, to identify them.

FACTORS GOVERNING THE FOREIGN DIRECT INVESTMENT DECISION: THE EARLIER PERIOD

The empirical evidence of factors motivating FDIUS should not be analyzed in isolation but, rather, should be examined within a theoretical framework. The following section briefly summarizes the foreign direct investment theories of the 1960s and 1970s, which were examined in detail in the previous chapter. Most were formulated in response to the challenges of explaining heavy U.S. foreign direct investment overseas during the 1950s and 1960s. An interesting question is whether empirical evidence relating to the reverse flow of such capital lends support to these early theories.

The Evaluation Of Foreign Direct Investment Theory: The Earlier Period

As indicated in Chapter II, most foreign direct investment theories, developed in the 1960s and 1970s, had microeconomic roots tied to the concept of monopolistic advantage. It was assumed that foreign direct investment took place because of market imperfections. Hymer (1976) and Kindelberger (1978, 1969) argued that multinationals are at a disadvantage with local firms who enjoy natural cost advantages based on location. To compete successfully, therefore, the multinational must possess an offsetting advantage typically in the form of a rent-yielding asset, such as a superior product or unique production know-how.[1]

Whereas some applied the theories of monopoly and monopolistic competition in explaining foreign direct investment flows, others found more evidence of oligopolistic behavior. Caves (1971), for example, concluded that firms in oligopolistic industries tend to expand their production operations overseas because, in doing so, they obtain intangible assets from their investments in advertising and in R&D. Accordingly, these firms create their own firm-specific, comparative advantage by producing and marketing differentiated products worldwide. Given the abhorrence of oligopolistic firms to wide swings in market shares, Knickerbocker (1974) and Graham (1974) argued that the timing and direction of multinational firm investments are largely determined by oligopolistic reactions to competitors' investments. A "follow the leader" pattern in the timing of multinational firm investments was cited as a governing factor.

The product life cycle theory of Vernon (1970) was an attempt to link comparative advantage in foreign direct investment to product differentiation. The different stages in the international life cycles of a product were traced as a firm exploited its advantage, first through exportation and then through foreign direct investment. Ultimately, product superiority was eroded through the transfer of technology and know-how to the foreign markets.

A broader interpretation of the firm's motivation to invest overseas was advanced by Buckley and Casson (1976). The argument was made that, for foreign investment to occur, there must be economies arising from the exploitation of market opportunities through the company's internal operations rather than through exporting or licensing activities. In short, there must exist some "internalization" advantage for the firm to serve the foreign market as a producer. According to Magee (1976, 1977), firms specializing in the production of technical knowledge invest overseas because valuable information can be transferred internally within organizations more efficiently and profitably than through external markets. For the technologically-sophisticated multinational, foreign direct investment is argued to be the best way to "appropriate" maximum returns on its investments in new knowledge and technologies.

The question naturally arises concerning the extent to which existing theories have been supported by empirical evidence. Probably the most common approach to examining motivation is the case study method, which requires undertaking and compiling analyses of particular direct investments by foreign enterprises in a search for generalizations. Short case studies are cited throughout the chapter; however, the focus is on more systematic approaches. Systematic studies tend to be of three different types. First, there are those which base conclusions on survey results, that is, on data derived primarily from interviews with executives of foreign multinational companies. A second technique is the cross-sectional statistical study in which the investigator seeks to identify the distinguishing characteristics of those firms which have undertaken foreign direct investment. The third type is the time-series statistical analysis, which involves finding macroeconomic variables that have been significantly related to foreign direct investment over the years.

Case Examples And Survey/Interview Data On Investment Motivation: The Earlier Period

Surveys indicate that managers of foreign multinational companies operating in the United States during the 1960s and 1970s perceived a variety of advantages gained through their investments. These advantages included: (a) gaining access to the large and growing U.S. market, (b) exploiting the low foreign exchange prices of U.S. property assets and obtaining lower-cost factors of production, (c) circumventing tariff and non-tariff barriers, (d) reducing transportation costs and delivery time, (e) obtaining property assets in a country that is relatively stable politically, (f) obtaining economies of scale by vertical integration and (g) obtaining knowledge of U.S. technology, management systems and marketing techniques and exploiting existing advantages based on superior process or product technology (Little 1978, p. 45).

Gaining Access to the U.S. Market

The size and growth of the U.S. market seemed to be a primary motivational factor during the early period of FDIUS, as documented in a variety of investment surveys. In 1971, Franko examined the business strategies and motivations of 38 Western European companies with direct investment interests in the U.S. and identified "market size and affluence" as the most frequently cited motivational factors (Franko 1971). In two additional surveys sponsored by the U.S. Department of Commerce (1976) and the New England Regional Commission (Mandell and Killian 1974], "gaining access to the large and growing U.S. market" was also cited as a primary investment motivation.

It is clear that the market access was perceived to be important by both foreign businessmen and by American observers. In the words of Robert Raeithel, the President of Rosenthal MetCeram, a German company operating in Rhode Island, "We were always on the lookout to get a foothold in the American market. If you want to be with it, you have to be in the country where your major customer is" (*The Boston Globe* 1992, p. 26). Furthermore, as articulated by Joseph Perella, a Boston investment banker, "A fundamental attraction is the size of the U.S. market. As foreign companies look at the way they have their assets spread around, many realize that they lack investments in a part of the world that has a significant portion of the world gross national product, namely, the United States" (The Wall Street Journal 1979a, p. 1).

However, an explanation of foreign direct investment flows based on market size and growth differentials alone is unsatisfactory for two reasons. First, there are obvious ways of gaining access to markets other than through direct investments, namely, through exporting and licensing arrangements. Second, one must cast doubt on the importance of this motivational factor since most firms surely seek profit rather than market share per se, and the profitability of a given investment is not necessarily related to market size.

The failure of companies to cite the profit motive or to refer specifically to "return on investment" have been addressed by Webley (1974). In examining survey results, he has concluded that businessmen are noticeably reticent about discussing the profit motive, a fact which is sufficient cause for casting doubt about the validity of the survey approach. Second, in reference to the lack of direct reference to ROI by interviewees on surveys, Webley cites the difficulty of estimating the expected rate-of-return. He doubts that foreign firms are motivated to invest in the United States because it "offers the maximum rate of return compared with alternative locations. It is more likely that an adequate rate of return is stipulated by those who determine the worldwide strategy of their corporations and thus other factors, such as defense of market share, become relatively important" (Webley 1974, p. 27).

Obtaining Lower-Cost Production Resources

Clearly, during the 1960s, the overvalued dollar discouraged foreign direct investment in the United States, and encouraged exportation, by making the foreign exchange prices of U.S. production resources artificially high and the dollar prices of foreign export products artificially low. In the 1970s, substantial depreciation of the dollar produced the opposite effect, and although surveys indicate that some companies are unwilling to base long-term investment decisions on what they view as short-term foreign exchange rate fluctuations, the length and extent of the dollar depreciation in the 1970s had a significant effect. Indeed, from 1971 to 1977 "the dollar depreciated 32 percent vis-a-vis the German mark, 44 percent vis-a-vis the Swiss franc, 27 percent vis-a-vis the Dutch guilder and 21 percent vis-a-vis the Japanese yen" (Little 1978, p. 46).

During this period, exchange rate considerations played a major role in inducing some companies to invest in the United States. For example, Ateljens des Charmilles, a Swiss machine tool company, decided to establish a manufacturing subsidiary in Rochester, New York, for a variety of reasons, the most important one being the company's deep concern over the appreciation of the Swiss franc vis-a-vis the dollar and the impact of that on the company's export market in the United States. Similarly, the Volvo group revealed that the company's vulnerability to changes in the exchange-rate relationship of the Swedish krona to the U.S. dollar was one of three major reasons behind the decision to establish a car assembly plant in Chesapeake, Virginia (Webley 1974, p. 24). According to W.T. Grimm & Co., a Chicago-based merger specialist, the weak dollar was a major factor behind the spurt in foreign acquisitions of U.S. companies from 162 in 1977 to 199 in 1978 (*The Wall Street Journal* 1979b, p. 1).

Obviously, the depreciation of the dollar during this period, in conjunction with depressed stock prices, reduced the relative cost of U.S. plants and equipment. Declines in relative labor costs also produced incentives to invest in America. United States Department of Labor indexes indicate that, between 1970 and 1976, unit labor costs measured in U.S. *dollars* grew more slowly in the United States than in the case of all other major industrial countries (Boner and Neif 1977, p. 16). The effect of the dollar's depreciation on relative labor costs was reinforced by wage inflation abroad, which generally outstripped the U.S. rate, and by social legislation and labor union agitation in foreign countries which translated into sharply rising fringe benefit packages for non-American workers.

Most foreign executives interviewed in the Department of Commerce study (1976) did not identify production cost criteria as decisive considerations in the decision to invest in the United States prior to 1974. However, in the mid- and late 1970s, unit labor cost trends became increasingly more important in attracting FDIUS.

Circumventing Tariff and Non-Tariff Barriers

Survey evidence is conclusive that multinational companies are motivated to circumvent tariff and non-tariff barriers by establishing direct investments abroad. In 1973, the U.S. Conference Board (1973) surveyed seventy-six chief executives of U.S. companies and asked each to identify the major reasons why firms invest overseas; sixty-four of the seventy-six executives surveyed mentioned tariff and non-tariff barriers as considerations.

In the Franco survey (1971), which was designed to examine the investment motives of thirty-eight European companies operating during the 1960s in the United States, respondents cited the existence (or threat) of U.S. tariffs, quotas and administrative regulations as one of the major motivational factors in the investment decision (Franco 1971, p. 27). In a similar vein, the comprehensive 1976 Department of Commerce study concluded that the existence of tariff and non-tariff barriers was of primary importance in inducing foreign direct investment in the United States during the 1970s. Indeed, the existence of U.S. trade restrictions was the motivational cost factor most frequently cited by respondents during the interview phase of the study.

Commerce Department survey results indicate that the American Selling Price System, used by the U.S. Government in duty evaluation, forced European chemical producers of benzoid-related substances to locate production facilities in the United States in the mid-1970s. It is not coincidental that the Japanese invested in assembly facilities for pickup trucks in the United States shortly after U.S. tariffs on truck imports were raised in 1971. Furthermore, there is strong evidence that U.S. initiated "orderly marketing arrangements" with the Japanese, involving U.S. pressure on Japan to voluntarily curtail exports, had the effect of attracting Japanese direct investment capital into the country. Finally, it is clear that increased U.S. tariffs on fabricated end-products during the 1970s convinced Canadian non-ferrous metal producers that the establishment of direct investments in the United States would be cost effective. However, survey results also indicate that U.S. trade restrictions are significant only when combined with other motivational factors, such as attractive local market opportunities. Tariffs alone, no matter how discriminatory against foreign exports, seem to be insufficient to attract the direct investor (U.S. Department of Commerce 1976, p. G-57).

Reducing Transportation Costs and Delivery Time

Rising transportation costs played a minor role in the 1960s and 1970s in inducing several foreign industries to substitute foreign direct investment for export activities in serving the U.S. market. The Department of Commerce survey indicated that reducing transportation costs became an important incentive in the case of bulk chemical products and heavy machinery (U.S. Department of Commerce, 1976, p. G-57). With reference to the latter, cost increases during the period became prohibitive factors for foreign firms contemplating initial export to the United States. Ateljens des Charmilles, the Swiss machine tool producer, cited "anticipated transportation cost and delivery time reductions" as two of the most important factors leading to the company's decision to construct a manufacturing facility in Rochester, New York (Webley 1974, p. 24). The company forecasted accurately that local manufacture and modular design would cut both transportation costs and delivery time substantially.

A similar forecast was made by Kikkoman, a Japanese soy sauce manufacturer, indicating that this incentive is relevant to foreign food processors as well. Kikkoman did invest in a manufacturing plant in the United States for the express purpose of reducing transportation costs (*The Wall Street Journal* 1977, p. 1). Of course, significant savings in transportation costs are not possible for all firms conducting assembly or manufacturing operations in the United States. Logically, this incentive is of little or no significance to companies exporting products to the United States which have a very high value per unit of weight or volume.

Obtaining Property Assets in a Country that is Relatively Stable

The political climate of the United States has been and continues to be one of the most favorable bases for business operations anywhere in the world according to survey data. Despite the atypical political disruptions in the United States during the 1960s and 1970s, the sociopolitical climate during this period still provided a strong added incentive to firms contemplating direct investments in the United States. Foreign executives interviewed by *Vision: The European Business Magazine* (Bradley and Lewis 1979) and by the U.S. Department of Commerce (1976) repeatedly referred to the United States as the "last bastion of free enterprise" or as the "last stronghold of Capitalism." Predictably, executives of foreign companies indicated that U.S. political stability is a necessary, but not a sufficient, condition for attracting foreign direct investment capital. It tends to be a decisive criterion only when combined with economic incentives, such as relatively low production costs.

Of course, the degree to which the political stability of the United States is viewed as attractive by the potential direct investor depends on the extent of perceived political deterioration in the home country. With respect to the latter, the Department of Commerce survey revealed considerable variation on a country-to-country basis. In the industrialized world, political deterioration during the 1960s and 1970s was perceived to have been most severe in the cases of the United Kingdom and the Netherlands. Some concern was expressed for Canada as an investment environment, considerably less for Germany and Switzerland, and at the other extreme end of the spectrum, investors viewed the Japanese political environment as a favorable factor (U.S. Department of Commerce 1976).

Concern about political deterioration was deepest, of course, in reference to third world countries. The failure of the North-South dialogue to resolve the key issues of conflict between governments of lesser developed countries and managements of multinational companies during the 1960s and 1970s is well documented. Although investments moved in the third world during this period, multinationals became increasingly more aware of their high risk exposure and vulnerability to rising nationalistic sentiments in poor countries. As a result, multinationals sought to lower their risk exposure by geographical diversification—a strategy which drew investment capital to the stable political environment of the United States.

Obtaining Economies of Scale by Vertical Integration

The availability of factors of production in the United States, unavailable or less available elsewhere, was cited in surveys in the 1960s and 1970s as a key factor in attracting a certain type of foreign direct investment capital (Little 1978, p. 45). The vertical integration of operations was cited specifically by Japanese and Canadian resource producers as a motive for exploiting direct investment opportunities in the United States. These firms were particularly interested in integrating backwards in order to maintain stable sources of required raw materials or intermediate goods and to obtain greater control over the cost of the same. The resources of concern included primarily oil, coal, metal, lumber, paper and certain basic agricultural inputs.

Of importance, but cited less frequently in surveys, was the desire by foreign producers to integrate forward for the purpose of expanding market shares and acquiring greater proportions of profits accruing to downstream customers (Little 1978; U.S. Department of Commerce 1976). In short, the large U.S. market provided an opportunity for foreign firms to capture scale economies through both backward and forward integration. It provided a learning experience as well.

Obtaining Knowledge of U.S. Technology, Management Systems Marketing Techniques and Exploiting Existing Advantages Based on Superior Product or Process Technology

The attractiveness of the United States as a direct investment market is certainly multifaceted. One factor often ignored is that foreign firms operating within the U.S. market have much to learn from competitive contacts with U.S. firms. This has been true historically. Modern management systems and innovative marketing techniques are traceable, for the most part, to the pioneering efforts of American firms. Furthermore, foreign firms in the United States have much to gain from the diffusion and dissemination of technology that occur within U.S. industries. However, although U.S. superiority in management, marketing and technology historically whetted the appetites of potential foreign direct investors, these same factors served effectively as real entry barriers. Obtaining knowledge from contact with competitors that employ superior technology, management systems and marketing techniques is possible only if the recipients survive in the marketplace. Obviously, in the face of competitive disadvantages in all these areas, the survival of inferior firms at early stages on their learning curves is doubtful.

In 1969, a number of European corporate executives were interviewed in a research project for the book, *Building the American-European Market*. The executives were asked about their firms' future plans for investing in the growing U.S. market. "Their degree of interest in this subject was close to zero" (Bradley and Lewis 1979, p. 44). In reference to the late 1960s, the study concluded that "for the management of any European firms, this decision to invest in the U.S. market presented real as well as psychological barriers" (Bradley and Lewis 1979, p. 79).

Ten years later, these attitudes changed dramatically, and indeed, there is strong evidence that those foreign executives interviewed in 1969 lacked powers of prognostication. During the 1960s and early 1970s, European firms successfully challenged U.S. multinational companies both in Europe and in the lesser developed world. The next logical step was worldwide expansion, including the extension of productive and marketing activities to North America. Many foreign firms, particularly from Europe and Japan, had by then acquired the technological, marketing, financial and management strength necessary to compete in the U.S. market. Indeed, by the 1970s, these foreign firms enjoyed technological superiority over U.S. firms in a number of areas, and surveys indicated that this was a motivational factor of major importance in spurring foreign direct investment (U.S. Department of Commerce 1976, p. G-39).

A national Academy of Engineering study reported, for example, that foreign pharmaceutical manufacturers in 1963 and 1964 introduced less than one-third of new chemical products, while in 1973 and 1974, by contrast, more than one-half of new chemical products reaching the market were of foreign origin. Numerous new technologies, both products and processes, were introduced into the U.S. market by British, German, French and Japanese firms (Little 1978, p. 46).

Most executives interviewed in the Department of Commerce study agreed that the combination of two key motivational factors were instrumental in their firms' decision to invest in the United States in the 1960s and 1970s: (a) the possession of competitive or superior product or process technology and (b) the large size of the U.S. market. A large market, per se, might be served through exportation; however, the executives concluded that it is both difficult and risky to launch innovative products in export markets. In such cases, direct foreign investment seems to be the more viable alternative. Furthermore, for the innovative foreign firm, market size is important because it permits firms to take advantage of scale economies which, in turn, supports expenditures on new product or process development. Finally, large scale production allows firms "to gain experience along a learning curve with respect to the conception, development and control of process technology which, over time, enables them to become even more successful innovators" (U.S. Department of Commerce 1976, p. G-39). In effect, technological superiority breeds foreign direct investment which, under the conditions cited above, breeds more innovation and continued competitive advantage.

Statistical Analyses Of Foreign Direct Investments In The United States: The Earlier Period

Relatively few econometric studies of foreign direct investment in the United States were conducted in the 1960s and 1970s, because attention was diverted to the larger outflow of investments made by U.S. multinationals overseas. Important examples include the Horst study (1972) which uncovered evidence that tariffs in Western Europe and Canada were significant in the 1960s in attracting USFDI and the Kwack study (1962) which found that U.S. investments overseas depended upon (1) the value of foreign output, (2) U.S. interest rates and (3) cash flow, net of dividends, of U.S. corporations during this period.

The Scaperlanda-Mauer model (1969), although originally designed to identify those basic factors which motivate U.S. direct investment overseas, was subsequently applied in an attempt to understand what attracted direct investment capital into the United States. In their original article, Scaperlanda and Maurer tested three hypotheses; namely, that U.S. direct foreign investment is motivated by (1) the size of the market in the host country, (2) economic growth and (3) tariff discrimination.[2] Empirical tests revealed that, regardless of the time period examined, only the size-of-market hypothesis could be supported statistically. This same model, employed by the staff of the Survey of Current Business in 1973, produced identical test results in reference to the flow of foreign direct investment into the United States from 1962 to 1971 (Leftwich 1973). The size of the U.S. market was the only variable statistically significant (at the one percent level) in all equations, and its coefficient always had the expected positive sign.

Although models, such as Scaperlanda-Mauer, offer adequate explanations of why host countries are attractive to potential investors, they fail to explain why the investment should originate in one foreign country rather than another or why the investment should not originate in the host country itself.[3] For example, why should market size per se in the United States be a significant inducement to foreign direct investment given the fact that domestic firms, with certain inherent cost advantages over foreign firms, are similarly attracted by a large national market? It is generally acknowledged and, indeed, confirmed by survey results that foreign firms must possess some distinct advantage over local firms, such as product differentiation, in order to offset the disadvantage of high entry costs, exchange risk and long lines of communication and control. Models, such as Scaperlanda-Mauer, fail to capture the nature of or the reasons for these "advantages" possessed by foreign firms.

Models of this type also fail to explain why foreign firms, attracted by favorable economic or political conditions in the home countries, choose to enter as producers rather than as exporters or licensors. The relative cost factors which must be weighed in selecting one of these three options are simply ignored. Also ignored are those determinants of aggregate domestic investment, such as expected profit, cost of capital, and so on, which are found in the better domestic investment demand models.[4]

Studies by Ajami and Ricks (1981) as well as by Franko (1976) revealed that foreign firms were attracted to the United States during this earlier period because of its market size, but not in isolation. Of greater significance was market size combined with the opportunity for firms to benefit from the market and R&D advantages of operating in the United States. Such advantages were shown to be industry specific, indicating that foreign firms had a propensity to move investment capital into research-intense and advertising-intense industries, motivated by a desire to benefit from high levels of U.S. technology and innovation.

This evidence is consistent with the findings of Vernon (1984), Flowers (1976) and Graham (1978). It also lends support to the results of the cross-sectional analysis of Caves (1974a, 1974b) and of Buckley and Dunning (1976). In examining foreign direct investment in Canada and Britain, these studies detected significant relationships between the extent of foreign direct investment in an industry and the importance of advertising expenditures, research and development expenditures, and professional labor skills in that industry. These results, argued Caves (1974a, 1974b), provide support for an "intangible asset" hypothesis associated with the profitable differentiation of the foreign firm's product. Caves' regressions also offer limited support of the importance of scale economies to multiplant enterprises as a factor in stimulating foreign direct investment.

An earlier time series analysis by Gruber, Mehta and Vernon (1976) supported the contention that foreign direct investment is a defensive measure by large oligopolies attempting to maintain foreign market shares in increasingly competitive environments. Flowers (1976) arrived at a similar conclusion by examining the oligopolistic reactions of large European and Canadian firms investing in the United States. Using regression analysis, Flowers found evidence that, in highly concentrated European and Canadian industries, firms tend to enter the United States in clusters of subsidiaries, apparently in response to the entry of the first investing firm from the particular industry in question.

A related but different follow-the-leader effect was identified by Graham (1974). Measuring cross-investments between U.S. and European multinationals, Graham observed an interesting investment pattern in which European investors in the United States seem to imitate the activities of U.S. investors in Europe. Increased U.S. direct investments in Europe seem to spur a reverse flow of investment capital after a relative short lag.

The Implication Of Empirical Findings: The Earlier Period

Surveys, combined with the compilation of case studies, have the advantage of directness and flexibility. The investigator can clearly focus on motivation and emphasize those features or questions that seem worth pursuing. Survey-based studies are by nature qualitative and impressionistic, in the sense that the investors' or respondents' evaluations offer only general notions of the relative importance of FDI motivation. Case studies and surveys also tend to be quite selective and may be rather costly (in time and money) to complete. Statistical approaches, on the other hand, provide quantitative criteria for establishing relative importance of varying causes and are rather easy to perform, given the data. The problem with this approach, of course, is obtaining relevant and accurate data, appropriate relationships, formulating and then interpreting the results. Subsequent results may only support, rather than confirm, particular theories.

These alternative approaches need not be mutually exclusive. Indeed, the theories being examined, and the availability (actual or potential) of information, basically determine the optimal approach. The product life cycle and most oligopolistic theories seem best addressed by various case studies. Theories emphasizing technological and product differentiation might best be evaluated in cross-sectional studies. The survey results described provide credibility for just about all of the hypotheses discussed.

In examining the experience of direct foreign investment in the United States in the 1960s and 1970s, some conflicting evidence about motivational factors did emerge, particularly in comparing survey results to some econometric studies. However, both surveys and statistical analyses indicate that the size of the U.S. market was a major factor inducing foreign businessmen to enter the country as producers. The importance of gaining knowledge of U.S. technology, management systems and marketing systems, while exploiting existing firm-specific advantages, was captured both in surveys and in more scientifically designed statistical analyses.

Significant empirical evidence did emerge from studies during the 1960s and 1970s in support of the notion that foreign direct investment belongs to the theory of imperfect competition, resting on a foundation of market imperfection. During this period, foreign firms were well aware of the natural cost disadvantages of operating subsidiaries in the United States in competition with U.S. firms, and were motivated to invest only where and when there were offsetting advantages based either on superior products, processes or resource endowments. Foreign firms, attracted by the large U.S. market, as well as by the opportunity to learn, invested to exploit their superiority and to capture scale economies.

Finally, it is clear from the evidence that some of the investment capital flowing into the United States in the 1960s and 1970s was rooted in traditional oligopolistic behavior. Investment motivations in these instances was reactive, not proactive, with visible follow-the-leader or defensive patterns.

FACTORS GOVERNING FOREIGN DIRECT INVESTMENT IN THE UNITED STATES: THE EXPERIENCE OF THE 1980s AND 1990s

Because good economic theory is designed to provide a simplified picture of complex reality, it is not surprising that the evolution of foreign direct investment theory in the 1980s was significantly influenced by changes in the nature and direction of international business activity during this period. The following section reviews briefly the most recent development in foreign direct investment theory, which are examined in detail in Chapter II.

The Evolution Of Foreign Direct Investment Theory: The Later Period

In the 1980s, direct foreign investment theory evolved to emphasize the transactional and the political aspects of international business. In reference to the former, the so-called "new" theories are essentially refinements of the theory of internalization (Rugman 1986) and eclectic theory (Dunning 1981). Such theories emphasize transaction cost market imperfections and have been referred to in the literature as the University of Reading School of Thought (Horaguchi and Toyne 1990). A second new direction in theory which emerged in the 1980s focuses on the exercise of power or influence underlying the direct foreign investment decision (Boddewyn 1988). This complex relationship is explored by use of political economy paradigms which examine the roles of all those social actions (including government) in the investment process that possess power, control or influence over relevant economic resources. Theories became macroeconomically oriented, in contrast to the earlier period, and some moved well beyond the normal boundaries of economics in examining the political, cultural and social factors motivating FDI.

In the 1980s, foreign direct investment theory also assumed more of a strategic and less of a static focus. Porter (1985, 1986) structured his theory around the strategic issue for established multinational rather than the processes by which firms become multinational. In the Porter model, firm strategies are examined, based on the extent to which cross border, value-adding investments are coordinated and on whether firm growth and expansion are centralized or decentralized.

Evolutionary theory also emerged in the 1980s, formulated on the view that the multinational firm is a social community whose productive knowledge defines a comparative advantage. This theory tends to view the multinational firm in a dynamic way as it evolves from its early origin to its global outreach, spanning international borders. The seminal works in the development of this theory were Nelson and Winter (1982) and Kogut and Lander (1992, 1993).

Finally, in the 1980s and 1990s, theory itself evolved to embrace the role of cultural similarities and differences in governing the location decisions of multinational firms. Davidson (1980) hypothesized that "firms initially exhibit strong preferences for near and similar cultures but exhibit little of the same with maturity." Johanson and Vahlne (1990) view the globalization of the firm as a process consisting of a series of small steps whereby firms expand operations as they gain experience in unfamiliar settings. Yu (1990) argues that two forms of experience are required for multinational growth: (1) country-specific experience and (2) general global operational experience. Cultural differences are said to govern the pace and location of a firm's growth in reference to the former.[5]

Empirical Evidence of Investment Motivation: The Later Period

Studies of the determinants of foreign direct investment in the United States conducted in the 1980s and early 1990s tend to be of two types. Because of the predictable lag in the empirical testing of theories, some studies of the past decade retest older theories (1960s and 1970s) and serve to either reaffirm or contradict earlier empirical findings cited above. Other studies are more at the frontier, testing theory as it has evolved in recent years. Both types are examined next.

The Ajami and Ricks study (1981) surveyed executives of recently established foreign owned multinationals. In support of earlier findings, these survey results revealed that the most frequently cited reason for foreign direct investment in the United States was "market size," but factor analysis in the same study supported the Sametz and Backman (1974) contention that foreign firms invested in the United States for the purpose of obtaining U.S. technology and know-how.

Some findings during this later period were contradictory. Lall and Siddharthan (1982) examined the industry distribution of foreign direct investment and found that industry-specific attributes such as R&D and advertising intensities were not statistically linked to the intensity of investment capital flowing into

U.S. manufacturing industries. On the other hand, Caves and Mehra (1986) and Kim and Lyn (1986) demonstrated that the dollar value of incoming foreign direct investment was indeed higher in R&D-intensive and advertising-intensive U.S. industries.

Based on more recent research, Kim and Nichols (1995) found evidence that foreign multinationals do concentrate on technology-intensive industries in their investment strategies and that industry-level characteristics, such as high levels of R&D intensity and capital intensity, are significant in influencing both the level and the direction of foreign direct investment. Similarly, Dunning and Narula (1995) found that the R&D intensity of foreign-owned production in the United States has been increasing over the past decade at a rate faster than that of indigenous firms. Multinationals from developing countries seem to be particularly active in this regard, seeking to diversify the location on their R&D activity and upgrade capabilities without surrendering resources to foreign control.[6]

In examining investment motivation in the 1980s and early 1990s, many of the same factors cited earlier in reference to the 1970s again surface. This is clear from a survey of foreign firms conducted by Sokoya and Tillery (1992). The following table lists the most frequently cited investment motives in descending order by the survey respondents (selected randomly from the Directory of Foreign Manufacturing in the United States).

Again, both the factors cited and the rank order of these factors support earlier studies, particularly those which identify the "large U.S. market" as the dominant magnet in attracting foreign direct investment capital.

A common thread that runs through the empirical literature linking the earlier and the later periods of this study is the identification of very visible, concrete advantages that foreign firms gain if they enter the U.S. market as producers. For example, M. van Niewkerk, chief of the balance-of-payments department of De Nederlandsche Bank N.V. in Amsterdam, attributes the investment attraction of the United States to what he calls "pulling" factors, namely:

> (1) The United States is the world's largest homogeneous market, offering a large and varied demand for high-quality products and services. (2) Science and technology are highly regarded, and U.S. society is generally

Table III.1. Reported Motives for Investing in the United States Importance in Descending Order

Extremely large U.S. market
Search for new markets
Need to be closer to consumers in order to give better service
General need for growth
General desire for growth
General desire for profits
Attractive political climate in the United States
Attractive United States attitude toward foreign investment
Desire to preserve markets that were established by exporting
Desire to compete abroad by employing technological, managerial, or financial advantages
Traditional U.S. receptivity to new products, methods, and ideas
U.S. Managerial and marketing know-how
Attractive U.S. capital markets
Desire for geographic dispersion as a means of spreading risks
Need for assured supplies or resources
Declining value of the U.S. Dollar, at the time of investment
To enjoy local incentives (such as taxes, credits, etc.)
Attractive U.S. supplies of important natural resources
Skill and efficiency of U.S. labor force
The possibility of import restrictions on primary products being imported to the United States\
Attractive U.S. technology
Desire to integrate forward and backward to reduce dependence on other firms
Desire to hold land or other foreign assets as a hedge against domestic inflation or as a store of value
Investments may be the only way to surmount risks
Comparatively less U.S. government control over business
Fear of furthur investments in home country because of political unrest

Source: Sokoya and Tillery (1992, p. 73)

> responsive to innovation. (3) Political, economic, and societal freedom guarantees foreign investors that they can operate on an equal footing with domestic producers. (4) The business atmosphere is flexible and dynamic; you can trim and integrate a company faster than you can in Europe, for instance. (5) Operating costs relative to many other countries are fairly small, and taxes are often lower than elsewhere. (6) The United States provides a relatively high rate of return on direct investment. (7) Depending on the strength of the dollar, the exchange rate can also provide a strong incentive (Strozier 1988, pp. 7-8).

Missing in this list of motivational factors, of course, are those organizational factors and market characteristics that translate into a firm-specific, comparative advantage for the investing company. Firms may be attracted by favorable economic and/or political conditions in the target country (pulling factors), but they must also be "pushed" by organizational and market factors that spell "success" in the competitive arena. Combined with the likelihood of profitable operations in the target area, because of some firm-specific advantage, may be the fear of long term market deterioration and profit erosion if the investment is not undertaken. More recent studies indicate that Japan's evolving overseas investment strategy in the 1980s viewed foreign direct investment in the United States as a means for capitalizing on existing opportunities and as a hedge against emerging threats to underlying corporate interests (Chernotsky 1987).

What are those organizational and market factors that have presented such an opportunity to Japanese and other foreign firms? Opportunities arise when the firm possesses superior managerial resources of some type. This may be in the form of a differentiated product (Hymer Thesis), intangible assets (Industry Structure Theory) or ownership-specific advantages (Internalization and Eclectic Theory). One implication of the foreign direct investment experience of the 1980s would seem to be that each of these theories in isolation is valid but underspecified.

Superiority in managerial resources can arise from a multitude of factors or conditions including administrative skills, technological sophistication (in reference to both product and process technology), marketing skills, organizational know-how and worker productivity. It is the firm's endowment of managerial resources that provides the "push" in the investment decision making process (Horaguchi and Toyne 1990, p. 487). The actual "endowment" composition will vary from firm to firm and from country to country. Studies of Japanese and Dutch investing companies, in particular, show a diversity in the managerial resource mix that produced profitable U.S. operations in the 1980s (Kimura 1989). This would clearly justify a broader, rather than a narrower, theoretical construct in explaining the investment pattern.

In the 1980s, size became increasingly more important in determining the direct investment activity of foreign firms; this included firm size, industry size and market size. Serving a growing market permitted foreign firms to expand operations, thereby capturing scale economies. In a 1991 study, Martin confirmed the importance of market structure (concentration and scale economies) and transaction costs in determining the extent of foreign direct investment in the United States. The rapid growth of the U.S. economy in the 1980s increased U.S. market size (a pull factor), and large foreign firms typically gained comparative advantage as producers in the United States in oligopolistic industries (Kahley 1987, p. 48).

Industry structure was clearly important in the foreign direct investment decision making process in the 1980s. Kimura's 1989 study confirmed some earlier empirical findings of Horst (1972), namely, that a firm's size is a persistent, powerful influence of the size of its foreign direct investment activity. Results suggest that the larger the domestic size of the firm, the larger the size of its foreign direct investment involvement. Kimura's results also suggest that a firm's domestic size in part reflects the strategic advantages that are also important in making it both large and profitable in the foreign location (Kimura 1989, p. 310). These strategic advantages were found to be linked to the broad breadth of product lines offered by the investing firm as well as to the internalized, vertical links that large firms were able to establish between the parent organizations and the overseas subsidiaries.

Empirical studies of the 1980s and 1990s indicate that factors motivating foreign direct investment are multidimensional and may vary from country to country and from industry to industry. Yu and Ito (1988), for example, found variations in motivational factors depending on the degree of competition in industries in host countries. Along the same lines, the Kim and Lyn study (1986) concluded that the motivations for investing in the United States differs, depending on the country of origin. Given the heterogeneous motivations that underlie foreign direct investment in the United States, the authors found it impossible to name a unifying theory broad enough to encompass all investment motivations. Sokoya and Tillery (1992) not only found significant motivational differences among firms investing in the United States from different countries, but also uncovered significant industry-to-industry motivational differences.

The macroeconomic determinants of FDIUS received very little attention in the 1960s and 1970s. A shift in emphasis took place in the empirical studies of the 1980s and 1990s. It is well-documented that U.S. macroeconomic policy (or the lack of the same in the case of the astronomical U.S. Federal deficits) had a profound effect on all aspects of U.S. balance-of-payments in the 1980s, including the foreign direct investment sub-account. For example, foreign exchange distortions caused by monetary and fiscal policy initiatives, as well as by central bank exchange market interventions, were very influential in the foreign direct investment decision-making process—a fact supported by survey data (Erich 1990). From 1985 to 1987 the U.S. dollar fell 80 percent against the Japanese yen and 67 percent against the German Deutsche mark.

> A U.S. company with a 1985 book value of $240 million, for example, had an effective price of $120 million to a Japanese buyer in 1987 because of the changing yen-dollar relationship. What's more, exchange rates also dropped the effective cost of insurance and other services (Erich 1990, p. 12).

Although survey data do reveal that FDI flows were influenced by exchange rate volatility during this period, statistical studies, attempting to nail down the nature of this relationship, produced both areas of agreement and areas of disagreement.[7] For example, Hultman and McGee (1988) demonstrated a significant correlation between changes in the U.S. real exchange rate and inward foreign direct investment during the 1980s. Their data seemed to suggest that the real exchange rate affects the relative cost of production as well as altering relative wealth across countries.

A more recent study (Klein and Rosengren 1994) confirms the strong correlation between U.S. inward FDI and the U.S. real exchange rate since the 1970s. However, where the Klein/Rosengren data consistently supported the significance of the relative "wealth" channel, it failed to support the relative "labor cost" channel. With reference to the "wealth" effect, Dewenter (1995), employing transaction-specific data of foreign acquisitions of U.S. properties during the 1975-1989 period, found no statistically significant relationship between the level of the exchange rate and FDI relative to domestic investment, after controlling for relative corporate wealth and the overall level of investment. In a study testing the effects that real exchange rate fluctuations had on FDIUS during the 1980s, Campa (1993) found exchange rate volatility to be negatively correlated with FDI in a sample of U.S. wholesale industries.

Finally, in a 1995 study, Goldberg and Kolstad found a significant correlation among FDI, export demand shocks and exchange rate variables. Variable real exchanges were shown to influence the location of multinational production facilities because of the export demand linkage. Because of a non-negative correlation existing between export demand and exchange rate shocks, multinationals seem to have the motivation to locate some of their productive capacity abroad.

Despite some areas of disagreement in the literature then, the relationship between the depreciating dollar of the mid-1980s and the heavy infusion of foreign direct investment during this period has been established. Nevertheless, it would be a mistake, according to Graham and Krugman (1995, pp. 38-42), to place much weight on exchange rate and current account developments over time in governing the growth of foreign direct investment. Well before the emergence of U.S. trade difficulties in the 1980s, foreign direct investment inflows had been expanding rapidly for an extended period of time. Viewing foreign direct investment exclusively within a foreign exchange market and balance-of-payments context is simply too limiting; it is more than merely a way of transferring capital between countries.

Graham and Krugman prefer to view foreign direct investment within a broader context. They suggest that in the early part of the post-World War II period, foreign direct investment was a one-way street. United States superiority on all fronts gave U.S. multinationals comparative advantage throughout the world. It is not surprising to Graham and Krugman, therefore, that this one-sided relationship between foreign multinationals and the United States should have changed, given the erosion in U.S. superiority on all fronts. "Instead of U.S. firms having a uniform advantage there are now many areas in which foreign-based firms have technological or managerial advantages that enable them to produce more effectively in the United States" (Graham and Krugman 1995, p. 41).

Although the root causes of the spurt in the inflow of foreign direct investment by the 1980s cannot be traced to deterioration in the U.S. current account, the Federal government's response to the problem did have an impact. The U.S. Congress, growing impatient with the large U.S. trade deficit of the 1980s, sponsored endless debates on "appropriate" policy approaches, culminating in the tough Omnibus Trade and Competitive Act of 1988. The bill contained tougher import restrictions and retaliatory clauses, but the mere threat of barriers during the debate was enough to spur a heavy influx of foreign direct investment capital during the late 1980s (Erich 1990, p. 13).

Whereas one could argue that it may not have been the explicit intention of the U.S. Federal government to attract foreign direct investment during this period, the same could not be said of state governments. One of the unique phenomena of the 1980s was the explosion of activity of states in seeking out foreign corporate investors. A U.S. government (NASDA) survey indicates that, by the late 1980s, three quarters of the states viewed foreign direct investment as an important part of their industrial development programs, and about half ranked it as very important or as a top priority (National Association of State Development Agencies 1986). Using a wide array of financial, tax and other incentives, states have been reaching out to foreign direct investors through direct mail, trade shows, foreign offices, print media, electronic advertising and video technology (Conway Data 1991). The outreach has been working, according to state reports; the trend continues into the 1990s. According to the May 27, 1991 edition of *Time* magazine, "virtually every state is going after a piece of the $400 billion worth of foreign investment in the U.S."

What are the enticements for the foreign firm contemplating a direct investment in the United States? In all 50 states the firm can expect investment tax credits or exemptions, loan assistance in the financing of land, buildings, machinery or equipment and trade support through strategically located foreign trade zones. In most states, the foreign firm will also be attracted by labor incentives such as state-provided recruitment and job training assistance, infrastructure programs in which government invests in social overhead capital and customized investment programs in which the company will receive significant tax and other benefits if it locates in specially-designated Enterprise Zones.

In short, it would be a gross oversimplification and distortion to call foreign direct investment in the United States today purely a market driven process. Government policies at both the

Federal and the state levels are clearly producing an investment diversion effect. An in-depth analysis of the nature, implications and effectiveness of U.S. government and state policy initiatives in this regard is conducted in Chapters V and VI respectively, following an examination of Japanese FDI in the United States in Chapter IV.

SUMMARY AND CONCLUSIONS

One lesson to be learned from an examination of foreign direct investment in the United States in the 1980s is that motivational factors are multidimensional and multifaceted. Advantageous conditions in the host country, for example, large market size, are part of the equation, but firms will not extend their operations overseas unless they possess some firm-specific advantage which can be translated into competitive advantage in the host country's marketplace. The advantage may be in the form of a superior product, process, manufacturing technique, marketing strategy, administrative skill, organizational know-how or worker productivity. It may be that the firm, because of some superior managerial resource, is best positioned to take advantage of a global presence through vertical, organizational linkages.

Certainly, there are a wide variety of economic factors that both push and pull the foreign direct investor. However, there are political, cultural and social factors as well that govern the flow of FDI.[8] The multitude of factors involved in the investment decision requires a multidisciplinary approach in examining the phenomenon. The evidence presented in this chapter clearly supports the evolution of FDI theory wherein the underlying analysis is elevated from the purely economic to more of a "politico-economic" and "socio-cultural" integrative approach.

Evidence on the strategic nature of multinational firm expansion also reveals that it is a mistake to view foreign direct investment as a decision made at a discrete point of time. The decision to enter a market is only the first step in the process. The dynamics of the entire process of operating a firm overseas must be incorporated into the design of a comprehensive FDI theory. As a firm evolves from its national origins to its global outreach, the motivational factors that drive the growth and expansion process evolve as well. While the dynamics of this process are not as yet well understood, multidisciplinary approaches to the study of these complex phenomena promise to shed light on what governs and influences investment decision making.

NOTES

1. In the literature, Hymer is credited for his pioneering work in tracing foreign direct investment systematically to its' microeconomic roots. For a short, but comprehensive summary of the Hymer thesis, see Kindleberger and Lindert (1978). Also see Hymer (1976) and Kindleberger (1969).
2. A complete description and evaluation of this model appear in Leftwich (1973, in sections IV.2 and IV.3 on pp. 30-34).
3. For a comprehensive and insightful review of the Scaperlanda-Mauer article, see Goldberg (1972).
4. It should be noted that Kwack does include these determinants in his model. See Goldberg (1972) and Kwack (1962).
5. For a thoughtful review of this literature, see Bonito and Gripsrid (1992).
6. Historically, most empirical studies testing the validity of foreign direct investment theory have focussed on FDI flows within the industrialized world. Recently, there has emerged in the literature interest in the nature and direction of FDI from multinationals in developing countries. Early empirical evidence seems to indicate that traditional FDI theory may not be a good fit. For example, in a study of Latin American and Caribbean direct investment in the United States, Krug and Daniels (1994) found little evidence to suggest that firms from these regions invest in the United States as the result of firm-specific or monopolistic advantages. Rather, evidence is offered suggesting that Latin American investment in the United States is largely motivated by capital flight. An unanswered question, of course, is why capital flight would take the form of direct, instead of portfolio, foreign investment.
7. For an interesting modeling approach for reducing the uncertainty associated with changes in foreign exchange rates when conducting economic evaluations of foreign direct investment decision making, see Lee and Sullivan (1995).
8. A comprehensive overview of all of these factors appears in the United Nations Centre on Transnational Corporations (1992).

Chapter IV

Japanese Foreign Direct Investment in the United States: A Special Case

INTRODUCTION

The decades of the 1970s, 1980s and 1990s have witnessed volatile swings in the flow of foreign direct investment into the United States. As indicated in Chapter I, FDI was essentially a one-way street until the 1970s. U.S. multinationals were active abroad as producers, but relatively few foreign multinationals ventured forth to establish operating subsidiaries in the United States. FDI became a two-way street in the 1970s, with a quantum jump in direct investment capital inflows. Levels remained high and rising throughout most of the 1980s until a turnaround occurred in the early 1990s, witnessed by sharply diminished inflows and even some disinvestments. In the mid-1990s, though, there is growing evidence of renewed interest on the part of foreign multinationals in FDIUS.

Swings in FDI inflows have been most pronounced in the case of Japan. The ratio of Japanese to total FDI in the United States increased dramatically in the 1970s and 1980s, while more recently it has declined, reflecting a retreat of Japanese direct investors from several sectors of the U.S. market. Following a statistical examination of the nature and quantitative dimensions of Japanese investment, the analysis in this chapter turns to motivational factors that have governed Japanese investment decisions over this time span. The heavy infusion of FDI in the United

States during the 1970s and 1980s produced a proliferation of studies of investment motivations in the economics literature. A review of this literature will be integrated into the chapter with a focus on those determinants of FDI in the United States that are uniquely relevant to Japanese multinationals.

In the literature, a conceptual framework for identifying foreign direct investment motivation has emerged. It is the central purpose of the chapter to present this framework and to examine its relevance and appropriateness in exploring the recent Japanese FDI experience in the United States. In examining investment motivations then, a contrast will be established between those factors that "pushed" and those that "pulled" the Japanese investments to U.S. shores. Also, the analysis will distinguish between those motivational factors which are market driven and those which seem to be linked more to political, cultural and social conditions. Finally, projections are made in the final section of the chapter based on current observable trends. An attempt will be made to identify new patterns and directions for Japanese foreign direct investment flows in the future.

THE QUANTITATIVE DIMENSIONS OF JAPANESE FDI IN THE UNITED STATES

Prior to the 1970s, Japanese FDIs in the United States were quantitatively insignificant. Capital inflows came mostly from Canada and Europe. The dominance of Canadian and European investments continued into the 1970s (Table IV.1 and IV.2), but this decade did witness a quantum jump in Japanese investment activity in the United States. Japan's share of total FDI in the United States was less than 2 percent at the beginning of the 1970s but had risen to over 6 percent by the end of this period.

In the late 1970s, Japanese investment activity accelerated sharply. In 1975, Japan's FDI position in the United States was $858 million. By 1980, this figure had risen to over $4 billion. As indicated by Tables IV.1 and IV.2, the acceleration of Japanese FDI in the United States, first witnessed in the 1970s, continued throughout the 1980s. Japanese investments increased not only in absolute terms but also in relation to capital inflows from other countries and regions. During this decade, the

Table IV.1. Foreign Direct Investment Position in the U.S.: Historical Cost Basis

	MILLIONS OF DOLLARS				*% OF ANNUAL CHANGE*			
Year	*All Areas*	*Canada*	*Europe*	*Japan*	*All Areas*	*Canada*	*Europe*	*Japan*
1973	18,284	4,244	12,504	259	[–]	[–]	[–]	[–]
1974	22,421	4,930	14,629	504	22.6	21.9	17.0	94.6
1975	26,740	5,416	16,533	858	19.3	4.4	13.0	70.2
1976	30,770	5,907	20,162	1,178	15.1	14.8	21.9	37.3
1977	34,595	5,660	23,754	1,755	12.4	–4.5	17.8	49.0
1978	40,831	6,166	27,895	2,688	18.0	9.1	17.4	53.2
1979	54,462	7,154	37,403	3,493	33.4	16.0	34.1	29.9
1980	68,351	10,074	45,731	4,225	25.5	40.8	22.2	21.0
1981	107,590	11,870	71,945	7,697	57.4	17.8	57.3	82.2
1982	123,590	11,435	82,767	9,697	14.9	–3.7	15.0	26.1
1983	135,313	11,115	92,481	11,336	9.5	–2.6	11.7	17.1
1984	164,563	15,286	108,211	16,044	21.6	37.5	17.0	41.5
1985	184,615	17,131	121,413	19,313	12.2	12.1	12.2	20.4
1986	220,414	20,318	144,181	26,824	19.4	18.6	18.8	38.9
1987	271,788	24,013	186,076	35,151	23.3	18.2	29.1	31.0
1988	328,850	27,361	216,418	53,354	21.0	13.9	16.3	51.8
1989	373,763	28,686	242,961	67,319	13.7	8.0	12.3	31.7
1990	403,735	27,733	256,496	83,498	8.0	–3.3	5.6	24.0
1991	419,108	36,835	256,053	95,142	3.8	32.8	–0.1	13.9
1992	427,566	37,843	255,570	99,628	2.1	2.7	–0.1	4.7
1993	464,110	40,143	287,084	99,208	8.6	6.1	12.3	–0.1
1994	504,401	43,223	312,876	103,120	8.7	7.7	9.0	3.9
1995	560,088	46,005	360,762	108,582	11.1	6.4	15.3	5.2
1996	630,045	53,845	410,425	118,116	12.5	17.0	13.8	8.8

Note: 1996 data are preliminary.

Source: U.S. Department of Commerce, *Survey of Current Business*, Various Issues

investment positions of all countries in the United States increased from $68.3 billion in 1980 to $373.8 billion by 1989. It was the period during which the global investments of multinational companies in general accelerated and foreign companies closed the gap established earlier by the dominance of U.S. multinationals.

Canadian and European FDI in the United States did increase in absolute terms during the 1980s (Table IV.1], but their relative shares of total investments eroded somewhat (Table IV.2). In the case of Europe, the degree of erosion was slight (from 66.9% share in 1980 to a 65% share in 1989). On the other hand, Canada's relative importance as a source of FDI capital for the United States eroded significantly from 14.7 percent share in 1980 to a 7.8 percent share in 1989

Table IV.2. Foreign Direct Investment Position in the U.S.: Historical Cost Basis Relative Shares

	Percentages of Total Investment		
	Canada	*Europe*	*Japan*
1973	22.1	68.4	1.4
1974	22.0	65.2	2.2
1975	19.2	61.8	3.2
1976	19.2	65.5	3.8
1977	16.3	68.7	5.1
1978	15.1	68.3	6.6.
1979	13.1	68.7	6.4
1980	14.7	66.9	6.2
1981	11.0	66.9	7.2
1982	9.3	67.0	7.8
1983	8.2	68.4	8.4
1984	9.3	65.8	9.8
1985	9.3	65.8	10.5
1986	9.2	65.4	12.2
1987	8.8	68.5	12.9
1988	8.3	65.8	16.2
1989	7.8	65.0	18.0
1990	6.9	63.5	20.7
1991	8.8	61.1	22.7
1992	8.9	60.0	23.3
1993	8.7	61.9	21.4
1994	8.6	62.0	20.4
1995	8.2	64.4	19.4
1996	8.5	65.1	18.7

Note: 1996 data are preliminary

Source: U.S. Department of Commerce, *Survey of Current Business*, Various Issues

(Table IV.2). Replacing Canada as a major capital exporter to the United States in the 1980s was, of course, Japan. From a relatively modest share of 6.2 percent in 1980, Japan by the end of the decade had became the second largest owner of U.S. property assets and the second largest foreign producer on U.S. soil, following Great Britain.

By 1990, Japan's share of the total FDI position of all foreign countries in the United States reached 20.7 percent. Despite some significant Japanese disinvestments in the early 1990s, Japan's relative importance as an owner and producer in the United States stabilized somewhat during this period. Its FDI share has remained in excess of 19 percent throughout the early and mid-1990s (Table IV.2), surpassing the aggregate investment position of Great Britain and all other countries in the world in 1992 and 1993.[1]

THE NATURE AND CHARACTERISTICS OF JAPANESE FDI IN THE UNITED STATES

Prior to World War II, all Japanese FDI in the United States served as an adjunct to international trade (Wilkins 1990). In this sense, it was single dimensional. The levels of Japanese investments over time reflected, therefore, the country's comparative advantages in international trade. Investments were undertaken by the general trading companies that carved out Japan's export and import trade and by the banks that financed the same (Caves 1976).

Following economic revival in the early post World War II years, Japanese trading companies and banks once again became the major players in the country's FDI activities. Japanese foreign investment in the United States and trade with the United States remained close complements until the early 1970s. The U.S. Department of Commerce estimated that in the early 1970s, U.S. affiliates of Japanese companies transacted about 94 percent of U.S. exports to Japan and approximately 86 percent of U.S. imports (Genther and Dalton 1990).

Some early Japanese investments were made in natural resource industries in the United States, motivated, of course, by the paucity of resource endowments in the homeland. Very little was invested in manufacturing industries. Japanese corporations simply did not possess the organizational strength or managerial prowess to compete internationally. The virtual absence of manufacturing activity was also a by-product of Japanese distrust of long-range foreign involvements and commitments, attitudes that changed in the late 1970s and 1980s (Chernotoky 1987).

As indicated earlier, Japan's share of total FDI in the United States was small prior to the mid-1970s, reflecting the fact that the country's trade advantages were limited as well. Acceleration in the country's FDI activities in the 1970s and 1980s, though, was correlated to the emergence of Japan as an import-dependent, export-driven nation. The course of Japan-U.S. international trade continued to drive the pace and direction of FDI activity to some extent (Caves 1993). However, in the 1980s, Japanese investment in the United States became more diversified and less linked to all the country's trade fortunes.

Historically, the largest share of Japan's direct investment has been in the area of "wholesale trade" (Tables IV.3 and IV.4]. The trade/investment link is clear here since wholesale trade involves the distribution system in the United States for imported products from Japan, such as automobiles. As revealed in Table IV.4, Japanese wholesale trade in the early 1980s accounted for approximately two-thirds of the country's total FDI in the United States. By way of contrast, in the early 1990s, only one-third of Japan's FDI position was wholesale trade related.

Another indicator of the closeness of the trade/investment link is the relative importance of banking activity. Again, it has been the historical role of Japanese banking subsidiaries abroad to support the country's trade activities. In the early 1980s, as indicated by Table IV.3 and IV.4, banking activity accounted for 12-15 percent of total Japanese FDI in the United States, but this relative share had dropped below 10 percent by the late 1980s and early/mid-1990s.

In combining the two categories, that is, wholesale trade and banking, the decline in trade-related Japanese FDI in the United States during the decade becomes more visible. For example, 1981 and 1982 wholesale trade and banking accounted for 80.1 percent and 77 percent of total Japanese FDI in the United States respectively. In 1991 and 1992, these relative shares had decreased to 41.2 percent and 42.7 percent respectively.

The decade of the 1980s also witnessed some important changes in the composition of Japanese FDI in the United States. Manufacturing investments grew in importance (Tables IV.3 and IV.4), continuing a trend that first became visible in the 1970s. A surge in real estate investments occurred in the mid- and late 1980s, reversed by the collapse of real estate prices in the United States in the late 1980s and early 1990s. In the mid- and late 1980s, Japanese investments became high-profile. Purchases of Rockefeller Center, Pebble Beach and Heavenly Valley were publicized worldwide. SONY came to Hollywood in 1989 with the purchase of Columbia Pictures, setting off a wave of takeovers in the U.S. entertainment industry by several Japanese publishing and electronic companies (Noble 1992).

Table IV.3. Japanese FDI in the United States by Industry Sector (Millions of Dollars)

	Manufacturing	*Wholesale Trade*	*Retail Trade*	*Banking*	*Finance*	*Insurance*	*Real Estate*
1981	1,321	4,985	67	1,180	[a]	[a]	305
1982	1,624	6,126	151	1,325	–570	169	396
1983	1,605	7,823	234	1,384	[a]	182	515
1984	2,460	9,689	252	1,853	513	138	744
1985	2,738	11,796	251	2,160	51	119	1,536
1986	3,578	13,687	290	2,704	2,087	[a]	2,941
1987	5,395	15,352	326	3,513	2,115	188	6,098
1988	12,222	18,390	346	3,895	2,863	[a]	10,017
1989	13,978	21,972	511	4,959	9,407	355	11,370
1990	17,145	26,148	632	5,970	8,873	388	15,245
1991	18,258	30,681	1,113	8,000	13,273	520	9,487
1992	18,321	32,841	980	8,809	13,087	486	9,909
1993	17,746	33,910	844	9,803	11,151	686	9,460
1994	18,691	35,691	1,141	10,223	12,778	778	9,773
1995	21,194	37,232	1,462	12,516	12,290	863	9,241
1996	29,454	38,021	1,741	6,816	21,322	771	8,823

Note: [a] Suppressed to avoid disclosure of data of industrial companies.
1996 data are preliminary.

Source: U.S. Department of Commerce, *Survey of Current Business,* Various Issues.

Table IV.4. Japanese FDI in the United States (By Industry Sector) Relative Shares

	Manufactu ring	*Wholesale Trade*	*Retail Trade*	*Banking*	*Finance*	*Insurance*	*Real Estate*	*Others*
1981	17.2	64.8	0.9	15.3	[a]	[a]	4.0	–22[b]
1982	16.8	63.3	1.6	13.7	–5.9	1.7	4.1	4.7
1983	14.1	69.0	2.1	12.2	[a]	[a]	4.5	–1.9[b]
1984	15.3	60.4	1.6	11.5	3.2	0.9	4.6	2.5
1985	14.1	61.1	1.3	11.2	0.3	0.6	7.9	3.5
1986	13.3	51.0	1.1	10.1	7.8	[a]	11.0	5.7
1987	15.2	43.7	0.9	10.0	6.0	0.5	17.3	6.4
1988	22.9	34.5	0.6	7.3	5.4	[a[	18.8	10.5
1989	20.8	32.6	0.7	7.4	14.0	0.5	16.9	7.1
1990	20.5	31.3	0.8	7.2	10.6	0.5	18.3	10.8
1991	19.5	32.7	1.2	8.5	14.2	0.6	10.1	13.2
1992	18.8	33.7	1.0	9.0	13.4	0.5	10.2	13.4
1993	18.4	35.2	0.9	10.2	11.6	0.7	9.8	13.2
1994	18.1	34.6	1.1	9.9	12.4	0.7	9.4	13.8
1995	19.5	34.3	1.3	11.5	11.3	0.8	8.5	12.8
1996	24.9	32.2	1.5	5.8	18.1	0.7	7.5	9.3

Note: [a] Suppressed to avoid disclosure of data of industrial companies.
[b] This negative value in the "other" category reflects, in large part, significant Japanese dis-investment in the U.S. petroleum industry.
1996 data are preliminary.

Source: U.S. Department of Commerce, *Survey of Current Business,* Various Issues.

Also, as revealed by Tables IV.3 and IV.4, the importance of Japanese non-bank, financial investments grew in the 1980s, not only in absolute terms, but also in relation to other types of FDI activities. Finally, Japanese investments in agricultural and natural resource industries in the United States increased significantly during the 1980s ("Others" category in Table IV.4), as land and other internal natural resource endowments in Japan became more restrictive in the face of a booming economy.

Some disinvestments have occurred in the early and mid 1990s, most notably in the area of real estate. The aforementioned collapse of U.S. real estate prices during the late 1980s, combined with more recent real estate deflation in Japan, has produced a persistent, negative investment psychology. The growth of investment in the financial sector in the 1990s has not matched that of the 1980s, but 1996 did witness a spurt in non-bank financial activity at the expense of the banking sector. Early evidence would seem to indicate that disintermediation has become part of recent Japanese FDI strategy (Tables IV.3 and IV.4).

Looking further, disaggregation of Japanese manufacturing investments in the United States (Table IV.5) reveals that investment tends to be concentrated in traditional low technology industries such as steel and automobiles, more than in high technology industries such as computers and semiconductors. However, the range of Japanese investments in manufacturing is wide, covering a broad spectrum of U.S. industrial activities.

The location of Japanese manufacturing facilities tends to be fairly broad based as well. As indicated by Table IV.6, Japanese subsidiaries by the end of the 1980s were operating manufacturing plants in 47 states in the United States. In terms of distribution of plants, however, there tends to be a concentration of these plants in California and in several states in the Southeast and Midwest.

Given the quantitative dimensions of Japanese FDIs in the United States, as well as the location and industrial characteristics, the focus of the discussion now turns to the motivating factors and conditions that have governed the pace, nature and directions of Japanese investments in the United States over the time period under study.

Table IV.5. Employment in U.S. Affiliates of Japanese Firms
Concentrations of Employment, by Industry

Industry & Products	1989 Employment	No. of Plants
Steel Works/Rolling Mills	32,727	17
Motor Vehicles	25,275	9
Rubber Tires	20,400	11
Computers & Peripherals	19,526	27
Color T.V.s & Audio Equipment	8,813	25
Heating & Cooling Equipment	7,361	17
Electronic Capacitors	7,042	12
Canned & Frozen Fish	6,850	37
Plastic Auto Parts/Casettes	6,383	40
Semiconductors	6,110	18
Motor Vehicle Parts	5,537	47
Printing-Business Forms	5,500	13
Printing Ink	5,361	48
Construction Equipment	4,550	9
Iron Castings	4,500	4
Printed Circuit Boards	4,475	10
Color TV Tubes	4,375	7
Auto Wire Harness	4,263	13
Computer Floppy/Compact Discs	4,032	16
Instruments-electrical Testing	3,925	9
Photographic & Copiers	3,847	15
Ball Bearings	3,838	15
Communications Equipment	3,581	13
Auto Seat Covers & Belts	2,813	10
Soft Drinks/Bottled Water	2,800	31
All Industries	**303,244**	**1,275**
Total U.S. Manufacturing	**19,612,000**	**98,000**

Source: Japan Economic Institute. Office of Industrial Trade and Office of Business Analysis Bureau of Labor Statistics, *Employment and Earnings*, March, 1990.

FACTORS MOTIVATING JAPANESE FDI: THE "PULL" OF THE U.S. MARKET

Traditionally, in identifying and examining factors that motivate foreign direct investment, determinants are subdivided into two major categories, that is: (1) those from the host country that "pull" investments and (2) those from the investing country that "push" companies to export capital. In the 1970s and 1980s, a unique blend of pull and push factors combined to motivate Japanese companies to accelerate their acquisitions

Table IV.6. Manufacturing Plants of Japanese Affiliates by State, 1989

State	*No. Of Plants*	*State*	*No. Of Plants*
California	235	Arkansas	13
Ohio	100	Maryland	12
Illinois	86	Oklahoma	11
Georgia	65	Arizona	10
Michigan	61	Colorado	10
Indiana	53	Connecticut	9
Kentucky	49	Nebraska	9
Texas	48	Maine	9
New Jersey	47	Nevada	9
North Carolina	43	Wisconsin	7
Tennessee	41	Mississippi	7
Washington	40	Iowa	6
Pennsylvania	36	Kansas	6
New York	35	Minnesota	5
Alaska	26	Vermont	4
Massachusetts	24	Utah	4
South Carolina	23	Louisiana	4
Oregon	22	Delware	2
New Hampshire	20	West Virginia	2
Alabama	17	New Mexico	2
Virginia	17	Rhode Island	1
Missouri	15	Idaho	1
Florida	14	South Dakota	1
Hawaii	14	Puerto Rico	5
	Total Plants: 1,275		

Source: Japan Economic Institute, Office of Industrial Trade, Office of Business Analysis

of U.S. property assets and to move more production and distribution activities to the U.S. market.

Market Size, Characteristics and Macroeconomic Conditions

On the "pull" side, it is clear that several aspects of the U.S. market per se served as magnets in attracting Japanese FDI during this period. The vast size of the U.S. market has been identified in several studies[2] as the most influential factor motivating Japanese and other multinational firms to enter the United States as producers. However, as pointed out by Andersson (1992), there were several positive aspects of the U.S. market system, beyond size, that combined to produce a very favorable investment climate. These characteristics included: (1) a strong commitment to private property rights and free enterprise, (2) political freedom and military security, (3) freedom for foreign owners to repatriate earnings, (4) access to ample supplies of raw materials, intermediate goods and skilled labor and (5) access to capital markets that are broad, deep and resilient.

Although favorable U.S. market characteristics in the United States during the 1970s and 1980s helped to explain the heavy inflow of foreign direct investment capital, in general these conditions in isolation are insufficient in explaining why Japanese FDIs rose in relation to all others. According to Chernotsky (1987), Japanese investment in the United States rose both in absolute and in relative terms during this period because of the degree of Japanese dependency on the U.S. market and because obstacles arose making it progressively more difficult for Japanese companies to serve the U.S. market as exporters.

By the mid-1970s, Japan relied on the U.S. market for more than one-fourth of its exports. However, its comparative advantage was in danger of erosion because of its deteriorating labor market and high labor costs relative to rates in the United States. The shift in emphasis toward serving the U.S. market more as a producer and less as an exporter had the effect of equalizing the cost of production, thereby enabling Japan to maintain market accessibility and competitiveness (Chernotsky 1987; Andersson 1992).

The abundance of raw materials suppliers was certainly a major factor that pulled Japanese FDI to the United States throughout the post-World War II period. However, as the country's competitive advantage in manufacturing expanded during the 1970s, internal natural resource limitations became more acutely felt and the motivation to invest in resource supplies overseas grew accordingly. As Japan's economy became driven by the need to exploit its competitive advantage in manufacturing, its multinationals became more driven to acquire direct access to raw material supplies in order to support that advantage. In a sense, therefore, Japanese FDIs in natural resource industries, initiated overseas during the 1970s and 1980s, became part of a broader global strategy to enhance the country's competitiveness worldwide by preserving existing advantages (Chernotsky 1987).

There were other advantages gained by Japanese firms in accessing the U.S. market in the 1970s and 1980s through FDIs. As indicated by Yoshide (1987a), setting up manufacturing operations in the United States promoted internal corporate efficiencies and flexibilities, as large Japanese companies transferred their products, processes, technologies and know-how directly to their subsidiaries. For smaller Japanese firms, setting up management and production operations in the United States facilitated the flow of materials, capital and goods to U.S. markets, helping to expand market presence.

During the 1970s and 1980s Japan was not satisfied with its market presence in the United States. It sought an expanding market share of this growing market. Through FDI, Japanese companies were able to establish closer customer relations, enabling them to respond more quickly to the changing customers' needs. This attention to customer wants and needs promoted appropriate changes in product design and development, enabling Japanese companies to adjust quickly to swings in market demand, thereby expanding market share. This location advantage, captured through FDI, could not be as readily achieved through exportation (Dillon 1989; Genther and Dalton 1990).

In the 1980s, a number of favorable macroeconomic conditions in the United States, beyond market size, were instrumental in attracting Japanese FDI. The combination of relatively high growth and moderate inflation created a healthy investment climate. Also, during this decade, huge U.S. federal deficits combined with the country's low savings rate to create a voracious U.S. appetite for overseas capital. Resulting monetary and interest rate conditions provided incentives for both foreign portfolio and direct investment. As the U.S. propensity to save declined during the 1980s, the propensity to consume, by definition, rose. The fact that U.S. economic growth was consumer-driven in the 1980s has been extensively documented. The steady eight year expansion from 1982 to 1990 in the United States was unprecedented in peace time and certainly provided rewards for Japanese and other multinationals that produced consumer products in the United States for the U.S. market (Bob 1990).

Trade Barriers: The Threat and The Reality

Of course, macroeconomic imbalance in the global economy can produce trade imbalance, which in turn can breed protectionism. A common strategy for circumventing trade restrictions that discriminate against exports is to locate within the protected market as a producer (Anderson 1988; Salvatore 1991). U.S. multinationals employed this strategy in Europe in response to E.C. trade restrictions in the 1960s and Japanese multinationals have employed the same strategy since the 1970s in protecting their market interests in the United States.

It is not coincidental that Japanese companies invested heavily in assembly facilities for pickup trucks in the United States shortly after U.S. tariffs on truck imports were raised in 1971. Similarly when "voluntary restraints" were imposed on exported Japanese color televisions in the late 1970s, Japanese television manufacturers moved some of their production operations to the United States. Not surprisingly, when voluntary export restraints were applied to the Japanese automobile industry in 1981, a similar direct investment response occurred. Toyota negotiated a joint venture with General Motors in California and Nissan and Honda invested in wholly owned assembly operations in Tennessee and Ohio respectively (Bob 1990).

Direct foreign investment inducement has been evident in other trade-sensitive areas as well. For example, in the late 1970s, anti-dumping investigations motivated Matsushita, Sharp, Sanyo and Toshiba to produce microwave ovens in the U.S. market (Chernotsky 1987). Additional evidence exists that trade tensions and the threat of harsher U.S. restrictions induced Japanese producers to move their operations to the United States in such key industries as steel (Kawasaki, Hitachi Metals and Nippon Kokan), consumer electronics (Hitachi and Matsushita) and semi-conductors (NEC).[3]

The link between U.S. trade barriers and Japanese FDI has been well established in the literature from evidence based on surveys, case studies and statistical analyses. In a statistical study that surveyed business executives of Japan's patent companies, Yoshida (1978a) confirmed that trade barriers played a major role in the decision of these companies to enter the United States as producers. The same conclusion was derived from the survey of 585 Japanese companies sponsored by the Japan Society (Bob 1990).

Salvatore (1991) found a significant correlation between increases in the number of U.S. non-tariff barriers and increased inflows of Japanese FDI in U.S. manufacturing. Takaoka in a 1991 study was able to document a close temporal connection between the rise of trade friction and ensuing decisions by Japanese companies to invest in the United States. Similarly, Encarnation (1986) demonstrated that the level of losses Japanese firms expect to incur from U.S. trade controls was significant in explaining the timing and sequencing of investments by Japanese auto manufacturers in the United States.

While several studies, cited above, confirm that Japanese FDI in the United States has been a defensive, reactionary measure employed to preserve market share, Bhagwati, Dinopoulas and Wong (1992) identify a second form of strategy, labeled a "quid pro quo" investment. This is a prohibitive measure (rather than a reactionary measure) taken by the investing company designed to avert protectionist sentiment or backlash. If Japanese companies succeed in serving the U.S. market more as internal producers and less as exporters, pressure on U.S. policy makers to restrict Japanese exports is alleviated.

The Effects of Exchange Rate Fluctuations

Japanese firms were "pulled" to the U.S. market as producers in the 1970s in part because of the opportunity to obtain lower-cost production resources. The weak dollar, which depreciated 21 percent vis-à-vis the Japanese yen from 1971 to 1977, played a major role in providing Japanese investors with cost incentives (Little 1984).

Apart from the early 1980s when the yen weakened in relationship to the soaring dollar, the past two decades have witnessed significant secular appreciation in the value of the yen. Over time, this currency strength has provided Japanese firms the equivalent of subsidies for their overseas asset and resource acquisitions. This subsidy has been particularly large in the case of Japanese direct investments in the United States (Chernotsky 1987).

Although there have been conflicting findings in the literature on the relationship between exchange rate changes in FDI, the weight of the evidence suggests that exchange rate movements do influence foreign direct investment decisions. Froot and Stein (1991) identified a wealth effect that shifts the demand for investment. As the dollar depreciates, the relative wealth of foreign investors increases, inducing them to invest more overseas. Since the dollar has depreciated more against the yen than against other major currencies in recent years, Japanese investors have not only been able to outbid American investors for U.S. resources but other foreign investors as well. Cushman (1988) found similar empirical evidence in support of Froot and Stein's "wealth effect."

There is also some indirect evidence that a relationship exists between FDI and exchange rate movement. Anderson (1988), concentrating on the effect of exchange rate fluctuations on wages, noted that the depreciation of the dollar since 1985, particularly vis-à-vis the yen, had significantly lowered U.S. wage rates relative to the foreign rates. The incentive for Japanese companies to take advantage of favorable wage conditions in the United States has been particularly strong.

Some research, though, questions whether short-term exchange rate fluctuations significantly influence FDI decisions that tend to rest more on long-term expectations. Bailey and Tavlas (1992) uncover only an ambiguous relationship between FDI in the United States and short-term exchange rate fluctuations. Ray (1991) questions the relative importance of exchange rate movements in this regard, arguing that foreign companies do not invest in the United States simply to take advantage of a cheap dollar. Survey data indicate that Japanese companies are influenced by favorable exchange rates in investing overseas, including in the United States, but that exchange rate considerations do rank relatively low in comparison to other motivating factors (Kim and Kim 1993).

Other aspects of this issue relate to foreign exchange risk. In a 1989 study, Ott argues that firms are motivated to substitute FDI for exports as a way to protect against exchange rate risk. Firms hold assets denominated in foreign currencies as a hedge against such risk and are able to reduce the burden of risk on production costs and sales prices by spreading production across several countries. In an empirical study, Cushman (1988) confirms the fact that multinationals are more likely to substitute FDI for exports when exchange risk increases. It appears that Japanese firms in the 1970s and 1980s, then, were influenced in their decisions to enter the United States as producers by favorable movements in the dollar price of yen and by the uncertainty created by exchange rate volatility.

The Quest To Obtain New Knowledge

Both the Japan Society survey (Bob 1990) and the Kim and Kim survey (1993) reveal that the advanced state of U.S. science and technology has attracted Japanese FDI to the United States. This was particularly true in the 1970s and early 1980s when Japanese companies were on a high point on the technology learning curve and sought to "ride down" that learning curve through its production presence in the U.S. market and through competitive contact with the U.S. based rivals.

Historically, technology in the global economy has primarily flowed from the homeland of foreign investors to markets in target countries. Interestingly, in the U.S./Japanese FDI relationship of the past two decades, the direction of technology transfer has been reversed. Japanese firms have been successful in capturing American technology, adapting it and translating it into competitive advantage in the marketplace. Success in this regard has served to whet the appetite of additional Japanese companies, motivating them to follow the same FDI path in their quest to obtain new technical knowledge and know-how in the U.S. market (Bob 1990).

It would be a mistake to assume that the Japanese learning experience from their presence in the United States as producers has been limited to the areas of science and technology, though. Japanese investments have not been restricted to technology-intensive companies. Historically, American firms have been pioneers in the development of modern management systems and innovative marketing techniques. Japanese firms have been motivated to come to the United States as producers in order to gain management insight and hone marketing skills, as well as acquire technical knowledge and know-how.

Public Policy: The Pull of Artificial Incentives

A foreign firm may be drawn to a target area by supportive government programs and practices as well as by favorable market characteristics. Although the U.S. Federal government does not have a formal incentive program designed to attract foreign investment, state and local governments have become quite aggressive in this regard. According to a survey conducted by the National Association of State Development Agencies (1986), nearly 75 percent of states in the mid-1980s considered FDI to be an integral part of their industrial development programs and over 50 percent ranked it as very important or as a top priority.

Some evidence exists in the literature confirming the effectiveness of artificial investment incentive programs. For example, offering potential investors tax abatements is a common incentive approach, and the link between tax levels and FDI levels has been established in several studies. Ott (1989), Maken (1988) and Tolchen and Tolchen (1988) demonstrate how shifts in tax rates may affect FDI decisions through their effect on expected rates of return and on profitability. Young (1988) found that the FDI decisions of "mature" firms are particularly sensitive to relative tax rates. Japanese firms that typically come to the United States as producers tend to be "mature" by definition.

The Kim and Kim survey (1993) reveals that state and local government incentive programs have captured the attention of Japanese multinationals. According to survey results, these incentive programs are cited as having "moderate" importance in influencing investment decision making. In the rank order of investment motives, they appear in the middle. In ranking the relative impact of specific programs, survey respondents cite the following in descending order: (1) infrastructure development support, (2) free training of workers, (3) tax abatements, (4) state-sponsored financial arrangements, (5) free land, water and electricity, and (6) administrative assistance.

This combination of incentives has been particularly effective in attracting Japanese manufacturing firms, such as auto companies, to Midwestern states (Newman and Rhee 1990). For nonmanufacturing investments, state sponsored incentive programs have been less successful.

FACTORS MOTIVATING JAPANESE FDI IN THE UNITED STATES: PUSH FACTORS

The analysis of FDI motivation now turns from "pull" to "push" factors. It is impossible to explain the quantum jump in Japanese

FDI in the United States during the 1970s and 1980s by focusing exclusive attention on favorable market, political, financial, social and cultural conditions in the United States. Conditions and transitions in Japanese industry and in the Japanese economy were also of importance in explaining the high propensity of Japanese firms to invest externally during this period.

Constraints on Corporate Expansion Within Japan

The heavy inflow of Japanese FDI into the United States in the 1970s can be explained, in large part, by the enormous gap in natural resource endowments between the two countries. The sectorial focus of Japanese FDI in the late 1970s was in mining, natural resource related investment and manufacturing (Andersson 1992). Favorable resource market conditions in the United States in the 1960s and 1970s, combined with overcrowding, environmental deterioration and rising production costs in Japan to provide strong incentives for Japanese firms to invest in the U.S. economy. Domestic natural resource limitations became progressively more acute as corporate activity expanded in Japan, fueled by dramatic advances in product and process technology. Although improvement in Japanese production efficiency helped to offset rising natural resource and labor costs, it became clear that the optimal production situation would be to employ advanced Japanese production technology in conjunction with relatively low U.S. resource costs in the U.S. marketplace. Foreign direct investment in the United States became the vehicle to achieve the best of all possible production worlds.

Interestingly, the gap in natural resource endowments between the United States and Japan became less important during the 1980s because of significant change in the composition of Japanese FDIs. The sectorial focus shifted from mining and manufacturing towards finance, real estate, commerce and services (Andersson 1992). Differentials in relative resource prices are, of course, less important in these latter sectors in determining comparative advantage. The explanation of why this transition took place in Japanese FDI extends beyond national differences in infrastructure, environment or resource endowments.

The Japanese Savings/Investment Imbalances

Direct foreign investment is a component of the long-term international flow of capital. Such capital flows are governed, in part, by the savings/investment balances in both investing and target countries. In this regard, macroeconomic forces in both Japan and the United States were important in the 1970s and 1980s in driving the heavy infusion of Japanese FDI into the United States.

Increases in net domestic savings in Japan during the 1970s and 1980s produced a supply side effect that led to increases in the Japanese propensity to invest externally, particularly in the United States. The gap between high Japanese personal savings rates and low U.S. rates has been well documented in the literature, but of equal importance during the 1970s and 1980s were general declines in Japanese governmental dissavings in the face of quantum jumps in U.S. Federal budgetary deficits, particularly in the 1980s (Caves 1993). Thus, in Japan, where private and public savings grew in excess of business and government investments domestically, "pushing" surplus funds overseas, the opposite occurred in the United States. The emerging creditor/debtor relationship between the two countries was clearly a by-product of significant national differences in the propensities to consume, save and invest privately as well as to tolerate deficit spending publicly.

How did these macroeconomics developments affect Japanese FDI in the final analysis? U.S. property assets became available at the right prices as U.S. companies sought revenue to pay down relatively high-cost debt obligations. For the Japanese multinational, macroeconomics developments had the effect of lowering the opportunity cost of capital for investments in the U.S. market.

Japanese Current Account Surpluses and the Public Policy Response

In a sense then, Japanese FDI was pushed overseas in the late 1960s, the 1970s and 1980s by Japanese government policy. In the face of rising current account surpluses, a blend of market and public policy reactions produced offsets in the flow of Japanese capital overseas. For instance, acceleration in Japanese FDI in the late 1960s coincided with the growth of payment surpluses and the accumulation of exchange reserves during this period (Chernotsky 1987). In the early 1970s, the surge continued, clearly stemming from the Japanese government's need to reduce burgeoning exchange reserves (Heller and Heller 1974).

By 1975, discontent over continued Japanese trade surpluses reached the boiling point. The Ministry of Finance of the Japanese government responded to the pressure by sponsoring a loan program designed to support the effort of Japanese corporations to set up overseas operations (Bob 1990). Other programs involving similar subsidies and incentives followed, motivated by the desire to diffuse the financial and emotional impact of high and rising Japanese current account surpluses.

In the early 1980s, the Japanese government relaxed regulations internally on financial intermediaries, permitting them to compete more aggressively for domestically-generated funds and to diversify their financial activities internationally. The effect was to promote the growth of both portfolio and direct investments overseas. There is evidence that these regulatory changes particularly spurred the sharp growth of Japanese FDI in the U.S. real estate and service-sector activity that occurred during the mid- and late 1980s (Tables IV.3 and IV.4).

Japanese FDI and the Exploitation of Competitive Advantage

Although the proper mix of macroeconomic conditions and government policy may motivate domestic firms to go overseas as producers, firms will not venture forth unless there is a competitive advantage to be exploited. Favorable macroeconomic conditions and supportive government policy may exist in a country such as Japan, promoting a capital surplus, but the macroeconomic environment per se is insufficient in explaining why surplus funds may exist at a particular point in time in the form of direct, rather than portfolio, investments. In explaining the reasons why firms opt for foreign direct investment, one must examine the microeconomic roots of the investment decision.

A review of foreign direct theories of the 1960s and 1970s reveals that the early architects of these FDI models linked investment activity to the existence of market imperfections[4]. In target countries, local firms have natural cost advantages over foreign firms based upon location. Thus, in a world of perfect competition and homogenous products, FDI would not take place. Given market imperfections, however, a foreign firm can compete successfully overseas if it possesses advantages capable of offsetting entry costs. Typically, such an advantage would be in the form of rent-yielding assets, such as a superior product or cost-effective production know-how.

It is clear that Japanese firms were "pulled" to the United States as producers in the 1970s and 1980s because of the size and attractive characteristics of the U.S. market. However, the U.S. market had been large and attractive during earlier periods (e.g., 1950s, early and mid-1960s) when very little Japanese FDI entered the country. Why the difference?

In the 1970s and 1980s, unlike earlier periods, Japanese product and production technology had advanced sufficiently in several key industries to produce the competitive edge necessary to induce firms to go overseas as producers. The Japanese FDI experience of this period lends support to those models of FDI behavior that are based on the theories of monopolistic competition and oligopoly. Japanese FDI successes in the United States have occurred in markets for differentiated products, primarily consumer goods, where Japanese superiority either on the product (revenue) side or the production (cost) side has given rise to quasi-rents. These rents have been captured through direct control of production by oligarchic groups of dominant Japanese firms in such key industries as automobiles, consumer electronics and photographic equipment (Georgiou and Weinhold 1992).

The Strategic Aspects of Japanese FDI in the United States

There has always been a strategic element in the movement of Japanese corporate interests and operations overseas. Early investments in resource-intensive industries were defensive in nature, being strategic responses to growing shortfalls in internally produced Japanese commodities. Later, investments in the areas of wholesale trade and manufacturing were designed to protect market shares in the U.S. Wholesale trade investments were complimentary, while supporting export trade directly, and manufacturing investments were substitutional, intended to offset any erosion in Japan's export market in the United States resulting from the threat or reality of prohibitive export barriers (Sarathy 1985).

According to the eclectic theory of foreign direct investment, a foreign firm must possess some form of strategic advantage over local firms in target countries to offset entry costs and the costs of operating in alien environments (Kimura 1989; Dunning 1988b; Buckley and Casson 1976). Strategic advantage is typically based on tangible or intangible assets held by investing firms that: (a) can be transferred from the home to the target country at low incremental costs and (b) can be sold and leased in target countries but only at relatively high transaction costs. In essence, investing firms are able to capture more rent by internalizing these assets in overseas production facilities than by selling or leasing them.

During the 1970s and 1980s, the Japanese success in transferring productive operations to the United States in the form of FDI was largely traceable to three broad categories of strategic assets. First, rapid advances in product technology and in accompanying know-how gave Japanese firms competitive edges in such manufacturing areas as color televisions, microwave ovens, integrated circuits, communications equipment, bearings and machine tools. Second, advances in Japanese production technology and techniques in heavy industries, such as automobiles, gave Japanese companies advantages in capturing rent through cost control and containment. Third, and perhaps most importantly, have been the strategic advantages that arise from the unique "organizational" characteristics of Japanese companies that have been transplanted to the U.S. markets and elsewhere (Chernotsky 1987).

What organizational characteristics of Japanese corporations are unique and relevant in this regard? In contrast to the U.S. firms, a high level of horizontal and vertical integration exists among Japanese corporate groups. Referred to in Japan as "keiretsu," these organizational arrangements are quite prevalent in heavy industries and in the financial sector. In Japan, integration is maintained on an ongoing basis through interlocking loan arrangements, personnel exchanges among member organizations, mutual stock holdings and even meetings/conferences of top level managers (Genther and Dalton 1990).

The keiretsu, as a strategic alliance, is the major form of interfirm cooperation and competition in Japan. Given the cultural, social and political environment of the country, it is also an efficient economic organization which minimizes transaction costs because of its effective positioning between markets and corporate hierarchies (Chang 1994). Most importantly, as an organizational arrangement among firms, it is exportable. Since the 1970s, Japanese companies have been successful in transferring their management systems and organizational arrangements overseas as part of their FDI strategy. The ability of Japanese companies to replicate their operating systems on U.S. soil has had a significant "push" factor, inducing these firms to move productive activities to their largest external market (Wassman and Yamamura 1989).

There is some evidence indicating that the keiretsu affiliation is a significant reason why Japanese automotive suppliers have invested in the United States over the past two decades. Although Hennart and Park (1994), investigating over 600 Japanese manufacturing firms in the mid-1980s, found no link between the direction of Japanese FDIs and keiretsu affiliations, their results have been questioned in the literature because of a failure to measure the mutual dependencies between core firms and their supplier organizations (Banerji and Sambharya 1996, p. 90). Using firm-level data for Japanese ancilliary automotive suppliers, Banerji and Sambharya (1996) were able to establish through logistical regression analysis that Japanese FDIs in the 1980s were influenced by keiretsu affiliations as well as by previous international experience and by firm size and dependence considerations.[5]

Within the Japanese keiretsu there is heavy reliance on trust and inter-firm cooperation. In horizontal keiretsu groupings, there are forms of mutual assistance and collaboration that produce economies and efficiencies. Specifically, these arrangements promote the horizontal transfer of valuable information and know-how, spur capital movements and reduce the risks associated with large ventures (Caves 1993). Similarly, vertical groupings of producers, suppliers and subcontractors serve to minimize transaction costs through the vertical integration of productive activity. This type of organizational arrangement has become an important tool in the promotion of the global sourcing strategies of Japanese multinationals.[6]

High levels of vertical integration are visible in the investment of Japanese automotive and electronics firms in the United States. This reflects the partial movement of existing industry keiretsu groups, such as Toyota and Sony, to the United States and to their efforts to duplicate subcontracting patterns followed in Japan with Japanese parts companies. This raises the issue, of course, of local content vs. import bias.

It is clear statistically that affiliates of Japanese multinationals had a higher propensity to import in the 1980s than in the case of affiliates from other countries.[7] For example, Japanese subsidiaries in 1987 imported approximately three times as much per worker as did other foreign-controlled subsidiaries (Graham and Krugman 1995, p. 77). To what extent was this high Japanese import propensity a byproduct of special supplier-customer relationships bred by the so-called vertical keiretsu? Logic dictates that a linkage must exist but there is some disagreement in the literature about its relative importance.

Unconvinced that the keiretsu system in the United States is systematically promoting Japanese affiliate imports and discriminating against local content, Graham and Krugman (1995) uncover other factors at work capable of producing this bias, including the relative inexperience of the Japanese as foreign direct investors. Inexperienced foreign affiliates, of course, with less opportunity to nurture supplier-customer relationships locally, are more likely to import from traditional, home-country suppliers. Thus, Graham and Krugman found it reasonable to expect that the Japanese would eventually follow the lead of European multinationals and increase the domestic content of their subsidiaries' output as they become more experienced as overseas producers.

In addition to industry keiretsu groups (both horizontal and vertical), in Japan there are financial organizations as well. The operations of financial keiretsu, groups of firms organized around a large bank, have also been transferred to the U.S. market. There are six financial keiretsu in Japan and the trading companies and main banks of all six currently operate in the United States (Genther and Dalton 1990). Table IV.7 reveals how many members of these financial organizations have affiliate operations in the United States.

It is clear from the evidence that both the industry and the financial keiretsu are capable of being transferred to overseas markets and that the Japanese in their FDI strategies have sought to capture the same economies and efficiencies externally as they have domestically. This reality remains a significant motivational factor in the global outreach of Japanese multinationals. It has been suggested that a combination of knowledge-based, productivity-based, and alliance-based strategies are responsible for the success of Japanese multinationals overseas (Smothers 1990). The successful transfer of industry and financial keiretsu overseas (particularly into the United States) has been a part of Japanese alliance-based strategy, complementing the knowledge-based and productivity-based strategies and helping to preserve their comparative advantage in global competition (Banerji and Sambharya 1996, p. 110).

From a strategic point of view, Japanese FDI in the United States and elsewhere are products of the country's industrial policy. It would be a serious mistake, however, to assume that the roots of this policy are traceable to post-World War II Japan only. The Japanese strategy of finding soft spots in foreign markets and in concentrating foreign subsidiaries in those areas where its comparative advantage is strongest is, in fact, deeply rooted in ancient Asian culture. Indeed, one can recognize the Japanese strategy for targeting and capturing foreign markets in the words of the ancient philosopher, Sun Tzu, whose classic treatise, *The Art of War* (1983), was written more than 2500 years ago.

> He will win who knows when to fight and when not to fight ... He will win who knows how to handle both superior and inferior forces... He will win who, prepared himself, waits to take the enemy unprepared.... If you know the enemy and know yourself, you need not fear the result of a hundred battles (Tzu 1983).

A LOOK AT THE 1990S AND BEYOND: CONCLUDING THOUGHTS

As indicated earlier, the decades of the 1970s and 1980s witnessed substantial, heavy net inflows of Japanese FDI capital into

Table VII. U.S. Manufacturing Presence of Financial Keiretsu Members

Keiretsu	*Members with U.S. Affiliates*	*Total Members in Japan*	*Percent of Tatal*
Sumitomo	9	10	90.0
Mitsubishi	13	20	65.0
Mitsui	7	11	63.6
Sanwa	18	29	62.1
Dai-chi	18	30	60.0
Fuyo	11	19	57.9

Note: Toyo Keizai, *Kigyo Keiretsu* Soran (1990), Office of Industrial Trade, Office of Business Analysis.

the United States. This chapter has examined a combination of factors and conditions that both pushed and pulled Japanese multinationals to U.S. shores and producers.

Japanese FDI data for the early 1990s tell a different story, however. Net inflows remained positive but diminished in 1991 and 1992. However, in 1993, for the first time in decades, there was a net outflow of Japanese FDI capital from the U.S. markets. Not surprisingly, the most quantitatively significant outflow occurred in the areas of real estate, finance and services in general.

Explanations for this reversal are not difficult to find. Again, factors were both of the push and pull variety. In the United States, for example, recession and lackluster recovery during the early 1990s impacted negatively on real growth, making foreign investments less attractive. Furthermore, the collapse of real estate prices, in combination with demand stagnation in other markets, reduced the profitability of existing Japanese investments in the United States, creating disincentives for new investments (Awanohara 1992).

There were supply factors at work in the Japanese economy as well. The heavy Japanese FDIs of the 1970s and 1980s were fueled by expanding capital availability. Bullish conditions in security markets combined with current account surpluses and favorable liquidity conditions, arising from banking deregulation, provided Japanese corporations with the financial support necessary to implement their FDI strategies.

Financial conditions in Japan have changed dramatically in the 1990s. Companies contemplating FDIs have been seriously constrained by a weak stock market performance, property price erosion, reduced corporate profits and higher capital costs (Quickel 1992; Thornton 1992). In the early 1990s, growth in Japan's FDI position in the United States was reduced by negative reinvested earnings, which arise when affiliates suffer losses or pay dividends to Japanese parent organizations in excess of earnings. In the category of "financial investments," net capital outflows were largely in the form of loans made by affiliates in the United States to cash strapped parents. Negative reinvested earnings became commonplace in real estate ventures as affiliates continued to pay dividends to parents despite negative earnings (U.S. Department of Commerce 1994).

Finally, it is evident that Japanese FDI in the United States declined in the early 1990s because of rising opportunity costs. Operating profit margins of Japanese owned ventures grew from 2 percent to 3 percent in Asia and from 0.8 percent to 1 percent in Europe from 1989 to 1991, but declined from 0.5 percent to minus 0.1 percent in the United States (Awanohara 1992). Because Japanese ventures in Europe and particularly in East Asia became more profitable than investments in North America, new Japanese investors turned away from the U.S. market.

Are the changes in the volume, nature and direction of Japanese FDIs of the early 1990s merely a temporary departure from trends of the 1980s or are they precursors of new Japanese global investment strategies? It is evident that certain economic developments that produced such change in the early 1990s will prove to be short-lived, but that others reflect new, long-term realities in the international economy.

In reference to the former, it is certainly true that during the early 1990s recessionary conditions in the United States and later in Japan distorted the foreign investment relationship between the two countries. However, FDI flows are governed more by long-term growth trends than by the vicissitudes of the business cycle. The size and growth of the U.S. economy were major attractions in the 1970s and 1980s, pulling Japanese companies to the United States as producers. It is predictable that if the real growth of the U.S. economy continues to be strong throughout the 1990s, the combination of rapid growth, large market size and opportunities to assimilate new technologies may prove to be irresistible once again to new Japanese direct investors.

In the early 1990s, Japanese disinvestments in the United States resulted, in part, from random developments, such as declines in U.S. real estate prices, which are clearly reversible. No one would predict long-term continuation of Japanese disinterest in U.S. property assets, based on the real estate price erosion of the late 1980s and early 1990s.

Similarly, Japanese disinvestments during that period resulted from a combination of financial development in Japan that: (1) limited the availability of venture capital to would-be foreign investors and (2) required companies already present in the U.S. market to restrict new funding to affiliates and even engage in net borrowings of affiliate revenues (U.S. Department of Commerce 1994). As indicated earlier, the pool of Japanese capital used historically to fund FDI dried up in the early 1990s as the result of bank loan defaults, corporate profit erosion and price collapses in stock and bond markets.

The mid-1990s did witness a spurt in Japanese FDIUS (Table IV.I) and optimists predicted more of the same, arguing that financial developments in Japan were random or cyclical, not secular. Even where problems were managerial in nature, such as inefficiencies in the organizational structure and operations of banks, optimists pointed to the change now taking place in Japan, designed to restore the viability of the system (Sapsford and Steiner 1995; Neff 1995).

It was argued that the Japanese recognized and were responding to the criticism that their banking industry had become too staid and resistant to change. The merger of the aggressive Mitsubishi Bank and the more conservative Bank of Tokyo, creating the largest private bank in the world, was cited as an example of a Japanese effort not only to capture economies through this ultimate form of horizontal integration, but also to infuse the banking system with more of an innovative and aggressive attitude. Japanese multinationals, comtemplating FDI in the future, were identified as the beneficiaries of this change (Sapsford 1995).

Unfortunately, the widespread Asian financial crises of 1997 and continued structural problems in the Japanese financial system have tempered this optimism. The future of Japan as a dominant foreign direct investor has become cloudy.

Prior to the Asian financial crisis, it was predicted that Japanese FDI would accelerate to 1980s levels, but that the United States' relative share would diminish. Despite the continued attractiveness of the United States as an investment target, it was predicted that Japanese FDI would become more evenly distributed geographically in the late 1990s. Positive macroeconomic trends in Asian countries and in other parts of the world would divert Japanese capital from traditional investment areas in the United States.

It is true that growing Japanese ties with other Asian nations were already visible in the mid-1990s (Neff 1995). Greater two-way flows of languages, entertainment and other forms of culture were providing a foundation capable of nurturing strong trade and investment relationships throughout Asia with Japanese corporations and banks well positioned to play dominant roles. The publication of a 1993 survey of Japanese corporations by the Japan Export/Import Bank revealed increased investor interest in third world target areas outside of Asia, particularly Latin America (American Embassy 1994). Again, favorable macroeconomic trends in Latin America were cited as potentially important "pull" factors. However, the survey was conducted before the collapse of the Mexican peso and the resulting financial instability has likely caused some rethinking in Japan and elsewhere concerning the overall attractiveness of third world markets, in general, and the Latin American market, in particular. The more recent Asian financial crisis has clearly compounded this uncertainity.

Given the uncertainties of how financial instability in Asia and in other parts of the world will affect the nature, pace and direction of international capital flows, it is impossible to predict what the precise FDI relationship between the United States and Japan will be as we move into the next century. It is predictable, however, that Japanese FDIUS in the foreseeable future will not closely resemble that of the 1980s.

NOTES

1. For a detailed statistical analysis of Japans's FDI position in the United States in the early 1990s, see U.S. Department of Commerce (1991, 1993a, 1994).

2. Studies that establish a link between foreign direct investment and market size in the target country, include U.S. Department of Commerce (1976), Ajami and Ricks (1981), Chernotsky (1987), Bob (1990) and Kim and Kim (1993).

3. There are several studies and reports documenting the role of trade restrictions and trade tensions between Japan and the United States in motivating Japanese companies to move their production facilities to U.S. shores. See Encarnation (1986), Yoshida (1987), Gentler and Dalton (1990), Salvatore (1991), Andersson (1992) and Bhagwati, Dinopoulas and Wong (1992).

4. Foreign direct investment theories developed in the 1960s and 1970s as logical extensions of traditional microeconomic analysis of product and factor markets. It was generally held that FDI takes place because of market imperfections under monopolistically competitive or oligopolistic market conditions. See Hymer (1976), Kindleberger (1969), Vernon (1970), Caves (1971), Dunning (1971), Graham (1974) and Tsurumi (1976).

5. Several other studies exist using firm-level and industry-level data for Japanese multinationals. Negandhi and Serapio (1992) edited a series of papers on trends and developments in Japanese FDI in the United States, which included some interesting comparative analyses of both Japanese manufacturing and service operations.

6. For good discussions of alternative sourcing strategies, see Buckley and Pearce (1979), Kotabe and Omura (1989) and Swamidass and Kotabe (1993).

7. For 1980s data on intra-firm trade between Japanese subsidiaries in the United States and their parent and other foreign groups, see Anderson and Noguchi (1995). Extensive sales data are used to shed light on the local content issue.

Chapter V

U.S. Government Policy and Inward Foreign Direct Investment

INTRODUCTION

The previous two chapters present empirical evidence in the identification of those factors or conditions in recent decades that have stimulated the inward flow of FDIUS, both in general (Chapter III) and in the special case of Japan (Chapter IV). Although evidence indicates that FDIUS is mostly market driven, justifying a focus on motivating factors emerging from private sector activity, public policy does play a secondary role.[1]

Chapters III and IV were designed to "touch all of the bases" in the attempt to identify all relevant factors that induce foreign firms to enter the U.S. as producers. However, the in-depth analysis in these chapters was reserved for private sector activity and market-driven motivational factors. The focus now turns to public policy initiatives and the role of government in attracting FDI. It is the purpose of this chapter to first examine the policies and practices of the Federal government in this regard, followed in Chapter VI with a parallel analysis of the role of state governments.

THE INFLUENCE OF PUBLIC POLICY: INTENDED OR UNINTENDED?

No one would argue against the assertion that the regulatory, monetary and fiscal powers of the United States or any other national government can be used effectively to influence both the pace and direction of inward FDI. Given the ability of the Federal government to create money, to engage in deficit spending constitutionally and to influence both interest rates and exchange rates, it is also irrefutable that the potential power of the Federal government to motivate inward FDI greatly exceeds that of state governments. Despite this, it is clear from the evidence that states have been much more active and aggressive than Washington in using public policy initiatives to attract FDI.[2] This is not to say that Federal policy has been irrelevant or unimportant in this regard, but rather that the impact has been generally indirect and mostly unintended.

Trade Policy and Investment Policy: Are They Linked?

The analysis in earlier chapters in this book established the fact that the threat and the reality of trade restrictions in recent U.S. history have motivated foreign firms to circumvent actual or expected trade barriers by substituting FDI for exports.[3] One could argue, therefore, that restrictive U.S. trade practices over the time period of this study have played a role in attracting FDI. Certainly, the presence of Japanese automobile manufacturing and assembly operations in the United States has been stimulated by the U.S. government's imposition of voluntary export controls on Japanese automotive exports in recent decades and by the ongoing threat of more of the same.

It can not be argued, however, that U.S. trade policy has been structured over time with the intention of substituting FDI inflows for imports. Recent U.S. trade policy has had the clearly articulated purpose of creating a more level playing field by promoting reciprocity and by seeking leverage in negotiations with trading partners through thinly-veiled or not-so-thinly-veiled threats of protection.[4]

Monetary, Fiscal and Exchange Rate Policy: FDI Implications

Exchange rate fluctuations influence FDI decision making.[5] Conceivably, monetary policy could be used to adjust exchange rates through an interest rate effect, thereby exerting an influence on international capital flows. However, there are two major reasons why the Federal Reserve has not targeted inward FDI through interest rate control. First, whereas foreign portfolio investment tends to be interest rate sensitive, foreign direct investment tends not to be. The financial objectives of a foreign purchaser of a domestic bond may be single-dimensional, but the objectives of the foreign purchaser of a domestic property asset tend to be multidimensional, not closely related to the current cost of money. It is true that FDI can be influenced by exchange rate fluctuations but, unless such movements are pronounced and prolonged, exchange rate effects tend to rank low in the list of factors that govern the FDI decision-making process.[6]

Second, apart from the question of effectiveness, the establishment of a certain level or rate of increase of inward FDI as a goal or target by the Federal Reserve would be inconsistent with the evolved philosophy of the central bank. Promoting domestic price stability has been the primary goal of U.S. monetary policy throughout the time period of this study, with secondary concern directed toward the domestic rate of unemployment, the real growth rate and the country's balance-of-payment position. Financial stability has also been established as a goal, but here the focus has been on short-term interest rate and exchange rate stability, designed to minimize the destabilizing effects of volatile short-term capital movements.[7] Even during the 1980s when the United States developed a voracious appetite for foreign capital and satisfied part of that appetite through inward FDI, it never became an objective of monetary policy or any other U.S. public policy to perpetuate the country's international capital account imbalance by intentionally inducing continued inflows of FDI.

This is not to say, however, that the heavy infusions of FDIUS during the 1980s were unrelated to U.S. government policy, but the fact remains that the effects were unintentional. In the late 1970s and early 1980s, the Federal Reserve waged an aggressive campaign against the serious inflation of the period by employing a policy of monetary contraction.[8] With growing Federal budgetary deficits during the early 1980s, the combination of fiscal expansion and monetary contraction drove U.S. interest rates up sharply, well above rates in foreign money markets. As foreign investments flowed into the United States seeking the higher rate of return, demand for the dollar rose with a predictable effect on the international value of the currency. The expensive dollar of the period drove the U.S. current account (on balance of payment) into deficit disequilibrium, offset, of course, by continued net capital inflows, including inward FDI.

The unintentional nature of this policy effect is clear. The tight money policy of the period was designed to fight inflation, not govern international capital flows. Similarly, the fiscal policy of the period had nothing to do with international balance-of-payment objectives but rather was based on the philosophical swing of the early Reagan administration to "supply-side" economics, involving tax cutting initiatives.[9]

On the surface, it might appear logical that the U.S. tax cuts of 1981 and the tax reform measures of 1986 would have had a significant impact on inward FDIUS, whether intended or not. However, the incentive effect of a tax cut may be quite different for a foreign firm, compared to a domestic firm, depending on the tax code of the particular foreign country in question.

In the case of MNCs from countries such as Canada and the Netherlands, which neither tax the earnings of overseas subsidiaries nor grant credits for taxes paid overseas, a tax cut in the United States should produce a positive investment incentive effect. On the other hand, for MNCs from countries that do both, such as Japan and the United Kingdom, the impact of a tax cut in the United States may be neutralized if tax credits are lost in the process. In short, the advantages of a lower tax liability in the United States could be offset or even more than offset by increased liabilities at home.[10]

Fiscal initiatives, of course, can be used aggressively with the intended effect of inducing MNCs to enter as foreign direct investors. It is common practice for countries to use tax incentives and subsidies to attract the investments of employment-creating foreign firms (OECD 1989, pp. 7-41). The United States is an exception to this common practice only insofar as the aggressive fiscal initiatives are undertaken by state governments, not by Washington (OECD 1995, pp. 49-52).

FDI POLICY OF THE U.S. GOVERNMENT

The U.S. government has not actively used its monetary or fiscal policy tools for the expressed purpose of enticing foreign multinationals to come to the United States as direct investors. This is not to say, however, that FDI policy in Washington is totally absent. The policy is one of treatment, not enticement, and the evolution of this policy over the time period of this study tends to be consistent with an ongoing philosophical commitment to neutrality. In short, the government's FDI policy reveals an historical bias neither in favor or opposed to foreign ownership of U.S. productive resources. This position of neutrality has been affirmed and reaffirmed in recent decades, most notably in 1977 and 1983 declarations by the Carter and Reagan Administrations respectively (Graham and Krugman 1995, p. 122).

Admittedly, the heavy infusion of FDIUS in the mid and late 1980s did generate some xenophobic reactions in the country, particularly directed at Japanese multinationals. This backlash translated into some lobbying pressure on both the Bush and Clinton administrations designed to induce Washington to restrict further inflows of FDI. These lobbying efforts were generally unsuccessful, however. Despite the fact that Federal policy in the 1990s is a bit more guarded than before, government has not retreated significantly from its broad philosophical commitment to neutrality (OECD 1995, pp. 31-33).

Neutrality, Openness and National Security

What is the essence of an FDI policy of neutrality? Government must be consistent in adhering to two principles. The first relates to the freedom to enter and expand. A neutral policy would disallow any government-imposed obstacles designed either to block the initial entry of the foreign multinational or to limit its freedom to move and grow. Second, a neutral policy involves a "level playing field" for established foreign direct investments. Government must not impose any special burden on the foreign company in comparison to its treatment of domestic firms (Graham and Krugman 1995, p. 122).

Historically, FDI policy of the U.S. government has abided by these general principles of neutrality in its openness and non-discriminatory treatment of foreign companies entering or already established on U.S. soil. In the early evolution of this policy, dating back to the late ninteenth and early twentieth centuries, exceptions to neutral treatment of foreign firms occurred only in cases of industries subject to Federal regulation, particularly those that contribute to national security.[11] Predictably, foreign investments continue to be restricted today in such security-sensitive industries as coastal shipping, air transport and atomic energy. U.S. policy is not unique in this regard inasmuch as most industrialized countries employ similar policies of restrictions in the name of national security (OECD 1995, p. 31).

A significant piece of legislation was passed in 1988 (with modifications in 1991 and 1992) as an amendment to the Defense Reduction Act of 1950. This so-called "Exon-Florio" provision gives the U.S. President the power to suspend or ban inward FDI (specifically mergers or takeovers) if it can be demonstrated that U.S. national security is threatened. Also divestiture of established foreign investments can be ordered by executive decree under the same provision.

Given the lack of clarity in the provision's definition of what comprises "national security," concern has been expressed in the United States and elsewhere about the possibility that U.S. officials may use Exon-Florio arbitrarily and capriciously to limit inward FDI. To date there is no evidence of such behavior. Indeed, of the approximately 900 cases reviewed between 1988 and 1994 under this provision, only one foreign company was denied entry (OECD 1995, p. 54).

Arbitrary and/or capricious treatment of foreign firms by U.S. officials would seem to be unlikely given the formal review process established by executive order in 1975. The reviewing mechanism is the Committee on Foreign Investment in the United States (CFIUS), composed of members from the Departments of Treasury, Commerce, Defense, Justice and State as well as from the Council of Economic Advisors, the Office of Management and Budget and the Office of the United States Trade Representative. The CFIUS is empowered by law to review the security implication of any foreign direct investment (Economic Policy Council 1991, p. 33). Although deliberations are confidential, the presence of representatives from so many agencies would seem to promote an interagency dialogue, at least exposing opposing positions of the different issues relating to FDI and national security.

Foreign companies continue to have limited entry into such Federally regulated industries as broadcasting and transportation, although exemptions can be granted by the appropriate Federal commissions. The link between these industries and national security are obvious; however, there has been a recent trend in Washington towards liberalization (Graham and Krugman 1995, pp. 122-123), perhaps spurred by the end of the cold war and the dismantling of the former Soviet Union.

FDI Policy and the Issue of Reciprocity

Although FDI policy in Washington today may be less shaped by national security concerns than in the recent past, the 1980s and 1990s have witnessed increased sensitivity to the issue of FDI reciprocity. This is not a new issue, but recent Federal legislation seems to reflect a strengthening of resolve of U.S. officials in this regard.

The receptivity of U.S. policy to foreign investors in such sectors as air transport, minerals mining, maritime transport, right-of-way for oil or gas pipelines and banking/financial services is conditional on the way U.S. investment interests in these specific sectors are treated abroad.[12] The seriousness of the U.S. government's intent to enforce reciprocity arrangements was reflected in the passage of section 301 of the Omnibus Trade and Competitiveness Act of 1988. Under its provision, the U.S. administration is authorized to take action against the nation's trade and investment partners that discriminate against U.S. foreign direct investments (OECD 1995, pp. 31-39).

Nevertheless, the intention of U.S. reciprocity provisions should not be misinterpreted nor should their importance be exaggerated. They are limited in scope to the aforementioned sectors and do not apply in general. Furthermore, they have not been designed to protect American industries from foreign competition; rather, the purpose is to apply appropriate pressure in order to open foreign markets to U.S. foreign direct investors. In a sense, therefore, recently-enacted U.S. reciprocity provisions in FDI law are not that inconsistent with the country's long-standing philosophical commitment to openness in its investment relationships with other countries. However, questions still arise about the appropriateness of the U.S. forcing the issue of reciprocity given its leadership role in the global economy. Questions also arise as to whether increased emphasis on reciprocity in FDI policy in the 1990s means that conditional national treatment has replaced unconditional treatment as the overriding principle in the U.S. foreign investment policy.

FDI Policy and the Promotion of U.S. Competitiveness

Deterioration in U.S. balance of payments, labor productivity and cost efficiency in U.S. industry all occurred during the 1980s. Accordingly, Congress passed several laws during this decade designed to promote more effective industry-government collaboration and to support private-sector technological development.[13] Since this increased emphasis on business-government cooperative efforts was designed to enhance the competitiveness of U.S. firms, the participation of foreign firms in these domestically-designed programs became an issue.

Several laws were passed in the 1990s designed to promote technological development in U.S. industry through private sector/public sector cooperation. Three laws in particular, the American Technology Preeminence Act of 1991, the Energy Policy Act of 1992, and the Technology Reinvestment Project under the Defense Appropriations Act of 1993 all permitted foreign firm participation, but only under reciprocity arrangements. Applications from foreign firms are approved only if it can be demonstrated that American companies have equal access, not only to the domestic programs per se, but also to comparable programs in the home countries of foreign applicants. Also, the participation of foreign firms in the three programs requires the demonstration of adequate and effective protection of U.S. intellectual property rights (OECD 1995, p. 41).

All firms, domestic or foreign, must show that participation in these Federal programs would be in the economic interest of the United States. However, considering the added burden of proof involved in the application process of foreign firms, domestic firms would seem to have a comparative advantage in gaining entry. The additional "hoops" that the foreign firm must jump through may not amount to prohibitive entry barriers, but they at least serve as psychological deterrents.

In the 1990s, new Federal legislation, promoting private sector—public sector collaboration in technology development, typically involved at the very least reciprocity requirements. Also, the updating of old legislation, such as the 1993 update of the National Cooperative Research Act, typically added the same requirements. Exceptions exist to the practice of imposing conditions on foreign participation in U.S. government sponsored programs, such as the High Performance Computing and Communications Program launched in 1991. The legislation establishing this particular program imposed no eligibility criteria (OECD 1995, p. 41), but this type of legislative neutrality, fairly commonplace in the 1980s and before, became quite atypical in the 1990s.

For decades, the United States played the role of leader in promoting an open-border, non-discriminating approach in the treatment of FDI. Historically, the United States has been uniquely open to the inflow, as well as the outflow, of FDI capital. In fact, it has been argued that the single most important non-macroeconomic factor explaining the rapid growth of FDI in the 1980s was the rapid dismantling of investment barriers in the world, which had the effect of spurring the international movement of capital (Economic Policy Council 1991, p. 16). As other countries, particularly in the industrialized world, began to emulate the openness of U.S. FDI policy, the results were predictable. U.S. leadership in the 1980s was at least partially responsible for the liberalization of international capital flows.

In the 1990s, the leadership role of the U.S. government in promoting free international capital movements is being questioned, however. Although the official U.S. policy remains one of neutrality and openness, particularly in reference to FDI, the U.S. government in the 1990s clearly has demonstrated increased sensitivity to the issue of equal competitive opportunity for U.S. firms abroad, including equal entry requirements for and treatment of U.S. multinationals. To American interest groups, reciprocity requirements may appear to be a necessary means of securing a "level playing field." To foreign interest groups, however, this may appear to be a subtle or not-so-subtle retreat of U.S. policy from historical positions of openness and neutrality.

The danger is that U.S. leadership in promoting capital mobility across international boundaries may be compromised by the foreign suspicion that beneath the rhetoric of reciprocity is a hidden agenda, namely, the promotion of U.S. competitive advantage in the global economy. Like Caesar's wife, it may be argued that the U.S. government should avoid even the appearance of an indiscretion, such as the introduction of a beggar-thy-neighbor industrial policy through the back door by the imposition of FDI restrictions.

FDI Policy and its Critics

Are the suspicions outlined above based on fiction or fact? Interestingly, the evidence is not clear in this regard. On one hand, critics of U.S. policy can not cite one major piece of legislation in the 1990s that blatantly violates the historical U.S. FDI policy of openness and neutrality. One may argue that there is an evolutionary process at work but, at least on the legislative side, there is no evidence of a "clean break" from historical policy.

On the other hand, if one examines the attitudes and policies of recent U.S. administrations, there is fuel to fan the flames of U.S. critics. Recent U.S. Presidents and their representatives have made it perfectly clear that the opening of foreign markets to U.S. investors and the fair treatment of U.S. multinationals abroad have become a priority and that U.S. government policy and practice will not always rely on the principle of unconditional national treatment (OECD 1995, p. 76).

Critics, of course, exist on both sides of the issue. As indicated earlier, there are those who believe that the United States has compromised its leadership position in the world by retreating from the principles of openness and neutrality in FDI policy. However, there are also critics on the other side who view U.S. policy as biased in favor of protecting foreign interests. Developments in the 1980s, particularly the heavy inflow of FDIUS, provided ammunition for the latter position.

Although concern in general was expressed in the 1980s about foreign acquisitions of U.S. property assets, most of the deep concern and alarm was directed at Japanese investments. This should not be surprising because, as indicated in the preceding chapter, Japanese FDIUS in the 1980s exceeded all others quantitatively and occurred in very visible sectors of the U.S. economy.[14]

In the extreme, Japan bashing in the 1980s took the form of accusing the Japanese not only of buying controlling interests in key U.S. industries, but of buying Washington as well. In his 1990 book, appropriately entitled *Agents of Influence*, Pat Choate painstakingly describes the strategy employed by Japanese lobbyists in currying the favor of Washington politicians, in influencing administrative and legislative decisions and in shaping U.S. government policy from the outside. What frightens Choate is not the effort but the apparent degree of success. He argues that "Japan's inside track allows its firms and organizations to short-circuit U.S. government decisions long before most people in the bureaucracy even know that something is afoot" (Choate 1990, p. 62). If one is to believe Choate, U.S. FDI policy has not deviated from its historical emphasis on openness. U.S. markets remain open to heavy infusions of FDI and favorable treatment of foreign firms, once established, continues because of supportive U.S. government policy. However, in Choate's view, government policy is supportive not because of any philosophical commitment in Washington to free international capital flows nor any conviction that FDI bestows benefits on the U.S. economy, but rather because permissive U.S. policy decisions have been unduly influenced by powerful foreign lobbyists, particularly the Japanese.

Developments in the 1990s would seem to contradict the notion of Japanese (or any other foreign) lobbying in Washington growing in power and influence to the point of being able to direct or redirect FDI policy. Foreign lobbyists would hardly be in favor of the changes that evolved in U.S. policy in the 1990s, emphasizing reciprocity, possible retaliation and the promotion of U.S. competitiveness in the world.

Other accusations and criticisms of U.S. FDI policy appeared in the press and the academic literature during the 1980s and early 1990s. Noteworthy among these was a controversial study entitled *Buying Into America* (Tolchin and Tolchin 1988). Two significant criticisms of U.S. policy, among others, were cited by the authors. First, U.S. officials were accused of focusing on the short-term benefits of foreign investment while, at the same time, ignoring the long-term effects, particularly the impact on foreign indebtedness. In reference to the decade of the 1980s, government policy in the view of the authors "took the form of non-policy, while inactivity, neglect and short-term patchwork solutions replaced any serious attempt to deal with the long-term consequences" (Tolchin and Tolchin 1988, p. 191).

Second, the accusation was made that rather than assuming a passive, neutral position, as is the conventional wisdom, the U.S. government in the 1980s aggressively searched for foreign capital in order to defray the country's mushrooming Federal deficits. In this campaign, Federal policy was said to "parallel state efforts in the indiscriminate search for foreign money" (Tolchin and Tolchin 1988, p. 194).

Two policy initiatives were cited to support this latter contention. The first was a change in the U.S. tax code under the 1984 Deficit Reduction Act that allowed foreigners who purchased American securities to eliminate the 30 percent withholding tax on interest earned and the second was a special Treasury security offering during the same period involving significant tax relief to foreign investors.

Accusations of this type were typical reactions to the unprecedented combination of macroeconomic developments that troubled the U.S. economy during the 1980s, including rising internal and external indebtedness. Although one can legitimately accuse the U.S. Federal government of public policy sins of omission and commission during this period, it is very questionable whether these problems or their consequences were linked at all to U.S. FDI policy.

In reference to the Tolchin and Tolchin arguments, it is rather farfetched to suggest a "parallel" between Federal and state efforts to attract foreign capital. Whereas most state governments have systematically and aggressively used formal incentives programs over the past three decades to attract FDI,[15] the Federal Government has not. The Federal tax measures of the mid 1980s, cited by Tolchin and Tolchin, do qualify as foreign investment incentives, but (a) they were clearly designed to attract foreign portfolio, not direct, capital and (b) they were isolated examples of government reacting randomly, not systematically, to short-term political pressure. Were these tax measures "short-term patchwork solutions?" This is a valid criticism, but it does not relate to FDI. There is no evidence that U.S. FDI policy has been shaped or even influenced by governmental "focusing on the short-term benefits of foreign investment."

Although U.S. FDI policy has evolved in the 1990s with growing government concerns about reciprocity and U.S. competitive advantage, the departure from the traditional philosophy of openness, neutrality and unconditional treatment has not been dramatic, certainly not on the legislative side. It may be a fair criticism to suggest that, because of evolving change in traditional FDI policies, the United States is at risk of compromising its leadership position as the world champion of free international capital flows. However, it is inaccurate to apply the criticism of "non-policy," "neglect" or "patchwork solutions" to the mainstream of U.S. policy on foreign direct investments.

FDI Policy and International Initiatives

U.S. FDI policy historically has not been formulated in geographical isolation. Policy has often been crafted through international negotiations and has taken form as treaties arising from successful negotiations. A good example is in the area of taxation. Over the years, the United States has signed treaties with over 40 countries, primarily designed to avoid the double taxation of income and to guarantee non-discriminatory tax treatment of United States and foreign companies operating abroad. In reference to inward FDIUS, there is an investment incentive factor inasmuch as some treaties "establish thresholds which permit foreign investors to carry on exploratory or preliminary business contacts in the United States without incurring income tax liability" (OECD 1995, p. 55).

U.S. participation in international agreements and treaties on economic policy is not new but, over the past two decades, the United States has been particularly active in seeking to extend economic cooperation to the area of international capital movements. Negotiations have taken place at different levels of government involving different parts of the world. For example, one priority dating back to the early 1980s has involved the negotiation of investment treaties with third world and Eastern European countries. The results have been fruitful inasmuch as negotiations have produced approximately thirty bilateral investment treaties between the United States and these countries. In general, the treaties called for "non-discrimination, free transfer of capital and returns, prompt, adequate and effective compensation, international arbitration, and discipline on performance requirements" (OECD 1995, p.48).

Noteworthy among recent bilateral agreements is the Canada-U.S. Free Trade Agreement. Precedent-setting clauses in this agreement include restrictions on the use of performance requirements imposed on foreign companies as well as dispute-settlement procedures governing the relationship between host countries and foreign companies (Graham and Krugman 1995, p. 170). The treaty deals with foreign investment issues broadly, obliging both countries to "non-discrimination against each other's companies with respect to new investment and acquisition and the conduct, operation, and sale of business enterprises and ownership shares" (Economic Policy Council 1991, p. 45).

Admittedly, most bilateral treaties involving the United States have been negotiated with developing countries. Agreements with other industrialized countries historically have been fewer in number and narrower in scope, usually limited to one industry or even a few firms (Economic Policy Council 1991, p. 45).

However, in the 1990s, the United States has become more active in seeking agreements with other industrialized countries on broader foreign investment issues. An example in this area is the recent Japan-United States Framework for a New Economic Partnership. Unlike earlier negotiations that focused almost exclusively on trade issues, this arrangement, mandating biannual meetings between the heads of state of the two countries, clearly establishes as a goal the removal of foreign investment impediments. The United States is unequivocal in seeking through these negotiations freer entry and fairer treatment for U.S. companies in Japan, not preferential treatment (OECD 1995, p. 49).

Although bilateral efforts to liberalize international capital flows continue in the 1990s, the focus has shifted somewhat to regional initiative and multilateral negotiations. On the regional level, NAFTA is a case in point. Historically, efforts to integrate regionally emphasized trade, not investment, issues. However, provisions for liberalizing direct investment took "stage center" during NAFTA negotiations. Accordingly, the ultimate agreement included "broad liberalization based on the principles of national treatment and non-discrimination, protection for foreign investors, dispute settlement procedures for both state-to-state and investor-to-state disputes, and in the case of Mexico, extensive liberalization" (OECD 1995, p. 48).

A second example of a regional initiative in the promotion of freer international capital flows is U.S. participation in the Asia-Pacific Economic Cooperation forum (APEC). In a 1993 summit meeting, member countries established foreign investment liberalization as one of the cornerstones of the APEC mission statement with an ongoing commitment to explore ways to nurture a more open environment for foreign investment within the region, including FDI.[16]

The United States has also participated in a leadership capacity in multilateral efforts sponsored both by the United Nations and by the OECD to establish rules for foreign direct investment. The focus in both cases has been the attempt to establish international codes of conduct defining the rights of host countries in regulating MNCs as well as the rights of MNCs in gaining market entry and non-discriminatory treatment. Both codes have been widely endorsed, at least in principle, by participating governments and by MNCs (Economic Policy Council 1991, p. 48).

Finally, the United States, once again in a leadership role, has been working in the 1990s within the context of the Uruguay Round negotiations of the General Agreement on Tariffs and Trade (GATT) to address some of the key issues of FDI, including national performance requirements. The Uruguay Round for the first time extended the purview of GATT to the service and investment sectors. Although the ultimate agreement did not address foreign investment impediments as fully as U.S. proponents of liberalization had called for, the Uruguay Round should be viewed as a major step in the right direction in broadening the focus of GATT beyond merchandise trade.[17]

CONCLUSIONS

It is true that U.S. macroeconomic policy has influenced and will continue to influence FDI flows. There is no compelling evidence, however, indicating that the U.S. government has used its monetary or fiscal powers to govern either inward of outward FDI proactively or systematically. Tax policy is a case in point. Unlike state governments that use tax write-offs and holidays for the obvious purpose of enticing foreign companies to locate within their borders, the Federal government historically has used tax policy to pursue domestic economic and social objectives. As indicated above, the Tax Reform Act of 1986 impacted FDI, but this was not a stated objective. The U.S. government does not use tax policy either to discriminate against foreign companies nor to intentionally seduce them to enter through the offering of subsidies.

U.S. interest rate or exchange rate policies may be formulated with one eye on short-term international capital flows, but, here again, there is no evidence that policy initiatives of this type have been used in recent decades to govern either inward or outward FDI.

This is not to suggest, of course, that the U.S. government operates without any FDI policy. Historically, this policy has been one of openness and neutrality, and the United States has been a leader among nations in seeking to liberalize practices and policies governing international capital flows.

Although the government's basic commitments to free international capital movements has remained unchanged, balance-of-payment pressures and other macroeconomic imbalances in the 1980s caused some dissent and dissatisfaction with the U.S. philosophy of investment openness. No significant piece of legislation in Washington during the late 1980s or 1990s signaled a significant retreat from that philosophy; however, Federal actions during this period, particularly on the administrative side, have reflected increased sensitivity to reciprocity issues and the competitive opportunities of U.S. multinationals operating abroad.

Seeking equal opportunities for U.S. companies is not new, but, during the 1990s, government policy and practice has become more aggressive in pursuing compliance under reciprocity requirements. Although few would argue that this reflects a fundamental change in the United States' commitment to investment openness and neutrality, it does appear to many as inappropriate for the world's historical champion of unconditional, non-discriminatory treatment of foreign investors.

The concern is not that the United States has imposed or may continue to impose restrictions on foreign companies for national security purposes. This is standard practice worldwide.

Rather, the concern is that the United States, under balance-of-payments or other pressures, may use reciprocity or "equal competitive opportunity" requirements to bring in a beggar-thy-neighbor industrial policy through the back door. To date, this perception is not reality but negative perceptions potentially could diminish the ability of the United States to play the role of leader in promoting freer international capital movements.

Is there evidence that the leadership role of the United States in this regard has eroded either as the result of the negative perceptions of others or waning U.S. interest or motivation? This seems not to be the case, although the nature of U.S. participation in efforts to promote a more open foreign investment climate globally has changed. U.S. sponsored bilateral efforts to liberalize FDI polices have continued into the 1990s, but recent efforts have become more regional and multilateral in scope. Unlike state government, the U.S. government does not use artificial incentives to attract FDI, but it does stimulate the inflow as well as the outflow of FDI through the negotiation of investment treaties or arrangements under regional initiatives such as NAFTA or APEC as well as through multilateral efforts under the direction of the OECD, GATT and the United Nations.

Is it likely that the protectionist pressures that surfaced in the United States in the 1980s and early 1990s will resurface in the foreseeable future, threatening the U.S. government's commitment to foreign investment liberalization through regional initiatives and multilateral efforts? Indeed, it is unlikely because the macroeconomic imbalances, particularly between the United States and Japan, that fueled the political tension in the 1980s and early 1990s, have significantly diminished. National governments more readily accept the international sharing of resources, including property assets, when foreign investment flows become more balanced.

NOTES

1. Survey data are presented in Chapter III, identifying motives reported by foreign firms for investing in the U.S. Public policy initiatives, such as the offering of tax holidays, are usually cited by respondents but typically rank low in terms of perceived importance. See Table III. 1 Japanese foreign direct investors tend to be a bit more receptive to investment incentive programs than the average foreign firm but still cite these programs as having only moderate importance in influencing their investment decision making. See Chapter IV.

2. The level and degree of state government activity in using investment incentive programs to attract FDI are explored in depth in Chapter VI.

3. For a complete discussion, see Chapters III and IV.

4. An excellent example of a U.S. policy initiative, using thinly-veiled blackmailing tactics, is the use of voluntary export restraints. For a cogent analysis of VERs and other non-tariff trade barriers, see Jones (1994).

5. The exchange rate influence on FDI is examined in Chapters III and IV.

6. This is clearly evident from Table III.1 in Chapter III.

7. The most authoritative source documenting the Federal Reserve focus on short-term interest rate and exchange rate stability is the monthly Federal Reserve Bulletin, particularly those issues containing the Board of Governors' "Monetary Policy Report to the Congress" and the report on "Treasury & Federal Reserve Operations." See Board of Governors (1994a, 1994b, 1996a, 1996b, 1995).

8. For an interesting, historical account of this monetarist experiment conducted by the Volcker Federal Reserve, see Neikirk (1987).

9. The Reagan desire to downsize government and apply supply-side economic policies were the motivating factors driving the fiscal initiatives of the period. Cogent, critical analyses of these initiatives can be found in Boskin (1987) and Klein (1983).

10. Despite the fact that the effects of taxation on international capital flows, including FDI, have attracted considerable attention in the economics literature, linkages and causal relationships remain unclear. For example, U.S. tax legislation in 1986, by reducing domestic tax credits and by eliminating special investment incentives, should have acted as a significant FDI disincentive, particularly for MNCs from the countries that permit their companies to write off foreign tax liabilities. The expected effect on FDIUS was not forthcoming, however. For a good discussion of this issue, see Graham and Krugman (1995). For examinations of related issues, see Janebra (1995), He and Li (1996), Porcano and Price (1996) and Swenson (1994).

11. A complete history of foreign investment in the United States, including a cogent examination of the early evolution of FDI policy, can be found in Wilkins (1989).

12. For a thorough examination of U.S. policies for these specific sectors, including enabling legislation, executive orders and other sources of authority, see OECD (1995, pp. 85-110).

13. The conventional wisdom holds that the U.S. government does not really have an industrial policy. Arguably, the U.S. legislative response to the macroeconomic problems and imbalances of the 1980s brought the United States as close to a formal industrial policy as any other historical occurrence. For a good discussion of this legislative response, see OECD (1995, pp. 39-52).

14. See Chapter IV.

15. This is one of the central findings in Chapter VI.

16. This agreement was reached at an APEC Ministerial Meeting on November 17-19, 1993, at Seattle, Washington. It is documented in the Meeting Proceedings. See APEC Ministers' Joint Statement (1993).

17. For a good analysis of the GATT Uruguay Round deliberations and agreements on foreign investment liberalization, see Schott (1994).

Chapter VI

State Incentive Programs and Foreign Direct Investment Motivation

Virtually every state is going after a piece of the $400 billion worth of foreign investment in the United States, and the fight is getting ugly (*Time Magazine*, May 27, 1991).

INTRODUCTION

The acceleration of foreign direct investment (FDI) flows over the past three decades has attracted considerable attention both in the international business management literature and in the economics literature. In the examination of investment motivation, market factors have been identified that both push and pull multinational companies to foreign shores as producers.[1]

However, there are non-market factors as well, most notably public policy initiatives, that influence both the pace and direction of FDI. In seeking to attract foreign capital, a host government may employ policy initiatives at two levels. At one level, macroeconomic policy may be used to indirectly attract FDIs by engineering favorable market effects. Specifically, fiscal, monetary or foreign exchange policy may be designed to affect changes in relative prices, incomes or interest rates for the purpose of making the macroeconomic environment of the host country more attractive.

Alternatively, the attempt by a host government to compete for FDI may be more direct. Artificial investment incentives, such as tax concessions may be granted, essentially offering the foreign investor a higher return than could be earned in the absence of such an incentive program.

The purpose of this chapter is to shed light on the nature and impact of public policy initiatives used by state governments in the United States to attract FDI. Although some reference will be made to the macroeconomic policies of the Federal government, basically to provide an analytical contrast, the focus will be on the investment incentive programs of the various state governments.

Where the Federal government has been relatively passive over time in seeking to govern FDI inflows and outflows through public policy initiatives, the states collectively have been very aggressive in this regard. The reasons are simple. At the national level there has been a heavy infusion of FDI capital into the United States over the past two decades because of market factors.[2] Given the large size and favorable characteristics of the U.S. market, no added incentives have been required in recent decades to induce foreign companies to move their production operations to U.S. shores.[3] However, once the decision is made to enter the U.S. market as a producer, the next question is one of location. Because of the economic and geographical diversity of the different regions of the United States, trade-offs exist, making locational decisions in the United States more difficult for foreign companies than is typically true of other host countries. U.S. states, enjoying considerable fiscal autonomy, see the foreign multinationals as a source of income and jobs and see the complexity of the location decision as a window of opportunity. Investment incentives are used to solidify strong locational advantages that some states may enjoy vis-à-vis rivals or to offset locational disadvantages in the case of others. Either way, the attempt is made to divert FDI artificially through competitive state incentive policies.

The analysis in this chapter is divided into five parts. First the locational characteristics of FDI in the United States are examined statistically with a focus on the state-by-state distribution. Second, the various investment incentive programs of the fifty states are described with an effort to provide an overview and to identify recent trends and practices. Following a review of the literature on the effectiveness of state investment incentive programs, new evidence is offered in the next section with an attempt to measure the degrees of success of such policy initiatives by linking locational decisions to the quantitative dimensions and qualities of incentive packages. Both primary and secondary data will be employed in support of arguments and conclusions. The final analytical section of the chapter attempts to ascertain whether domestic and foreign corporations have different locational preferences in the United States and whether they react to state investment incentive programs differently.

THE GEOGRAPHICAL DISTRIBUTION OF FDI IN THE UNITED STATES (BY STATE)

Evidence indicates that foreign companies locating their production facilities in the United States in the 1970s and early 1980s favored the large industrial states, such as California, New York, Texas, New Jersey and Illinois. Table VI.1 shows the value of foreign-owned property, plants and equipment accumulated state-by-state as of 1983 and also the direct employment effect. Since FDI during this period tended to be manufacturing-based, it is not surprising that foreign companies tended to locate their operation in resource-rich, high-income and large-market states.

This conclusion was generally supported by the findings of Ondrich and Wasylenko (1993) in their comprehensive study of the location of new foreign manufacturing plants in the United States from 1978 to 1987. However, two significant shifts in the locational distribution of FDIs were observed in the mid- and late 1980s.

First, although foreign companies, moving their productive operations to the United States, continued to favor the larger and more industrialized states such as New York and California, there was a discernible swing to the eastern sector of the United States. Second, among the states attracting the largest shares of new FDI capital during this period were southeastern states such as North Carolina, Georgia and Tennessee (Figure VI.1). This is interesting because these three states do not rank at the top of the list in terms of market size or resource wealth.

From 1978 to 1987 there was a heavy concentration of new FDI in manufacturing in seven states: New York, Illinois, California, Texas, Tennessee, Georgia, and North Carolina. The degree of locational concentration during this period is reflected by the fact that these seven states accounted for nearly fifty percent of total FDI (Ondrich and Wasylenko 1993, p. 40).

In terms of country of origin, there was another type of FDI concentration during the late 1970s and 1980s. Nearly three-

Table VI.1. *Foreign Direct Investment in The United States, 1983*

	Value of Foreign-owned Property, Plants & Equipment (in millions)	*Direct Jobs Resulting from Foreign Investment*		*Value of Foreign-owned Property, Plants & Equipment (in millions)*	*Direct Jobs Resulting from Foreign Investment*
Alabama	3,313	30,711	Nebraska	335	5,605
Alaska	11,636	6,787	Nevada	676	5,593
Arizona	3,459	25,828	New Hampshire	461	15,183
Arkansas	811	17,112	New Jersey	7,886	135,507
California	27,114	253,596	New Mexico	1,342	27,750
Colorado	3,821	29,701	New York	11,397	217,150
Connecticut	1,624	40,135	North Carolina	5,977	95,353
Delaware	2,290	34,688	North Dakota	1,542	3,560
Florida	8,138	84,568	Ohio	7,028	121,667
Georgia	5,979	86,014	Oklahoma	4,056	27,821
D. C.	921	4,095	Oregon	907	13,214
Hawaii	1,599	16,198	Pennsylvania	6,926	129,818
Idaho	515	4,512	Puerto Rico	432	8,969
Illinois	6,967	125,808	Rhode Island	414	11,198
Indiana	1,968	47,269	South Carolina	5,468	59,875
Iowa	1,195	19,172	South Dakota	371	1,987
Kansas	997	14,106	Tennessee	4,707	55,818
Kentucky	2,780	28,750	Texas	31,047	192,787
Louisiana	10,417	49,396	Utah	2,409	13,776
Maine	1,920	21,675	Vermont	367	5,635
Maryland	2,536	44,762	Virginia	3,972	56,859
Massachusetts	2,524	66,098	Washington	2,864	28,646
Michigan	4,825	66,745	West Virginia	5,212	33,179
Minnesota	3,741	30,137	Wisconsin	2,772	63,797
Mississippi	1,851	13,515	Wyoming	2,232	3,833
Missouri	2,443	37,191			
Montona	1,904	3,153	**TOTAL**	**241,600**	**2,526,183**

Source: Clarke (1986, p. 81); data derived from the U.S. Department of Commerce, Bureau of Economic Analysis.

quarters of all new FDI in manufacturing plants during this period were established by multinationals from four countries: Japan, Germany, the United Kingdom and Canada.

Figures VI.2 through VI.5 indicate the locational preferences of foreign firms from these countries. During this period, Japanese investors favored large-market, high income states in their

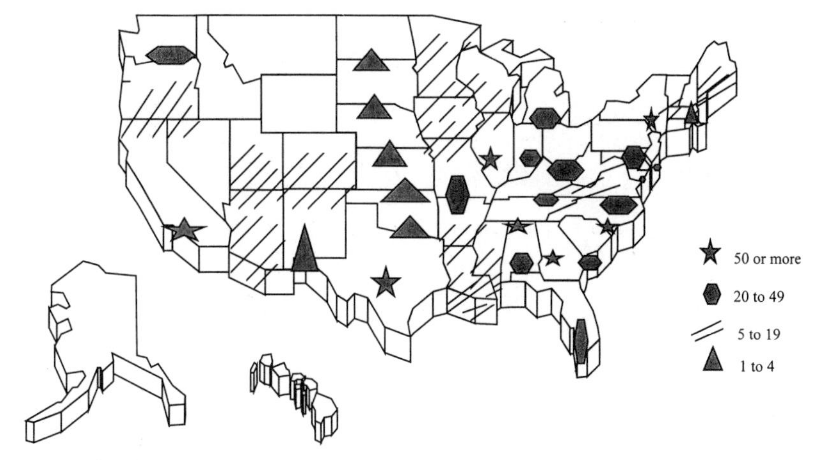

Source: Ondrich & Wasylenko (1993, p. 41).

Figure VI.1. Location of Foreign Manufacturing Plants in the United States, 1978 to 1987: 1,197 Total

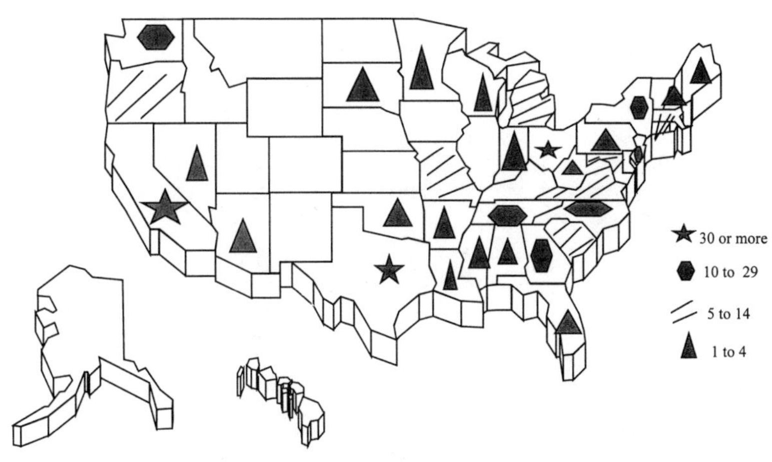

Source: Ondrich & Wasylenko (1993, p. 44).

Figure VI.2. Location of Japanese Manufacturing Plants in the United States, 1978 to 1987: 402 Total

locational strategies and there tended to be less of a regional emphasis in the placement of their manufacturing plants. The Japanese invested heavily in the West, the Southeast and Northeast and the Midwest (Figure VI.2).

By way of contrast, German and British investments in manufacturing tend to be more concentrated in the eastern part of the

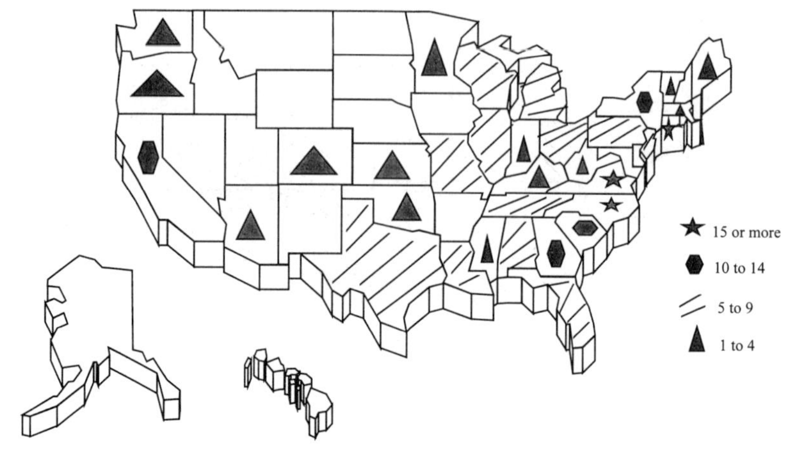

Source: Ondrich & Wasylenko (1993, p. 43).

Figure VI.3. Location of German Manufacturing Plants in the United States, 1978 to 1987: 215 Total

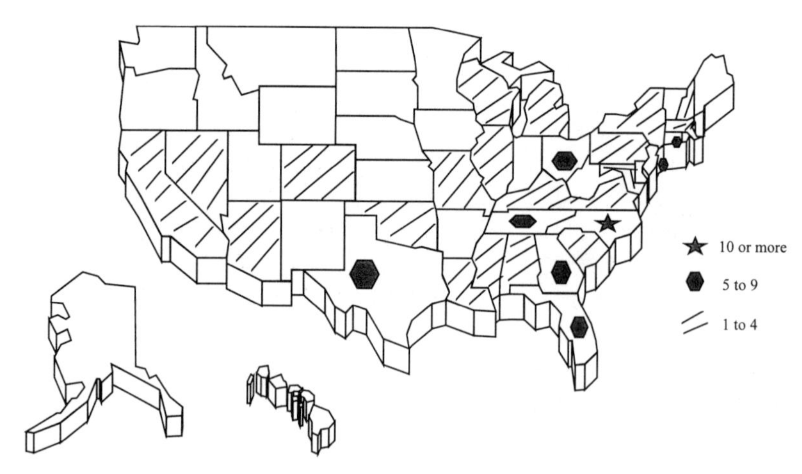

Source: Ondrich & Wasylenko (1993, p. 46).

Figure VI.4. Location of United Kingdom Manufacturing Plants in the United States, 1978 to 1987: 123 Total

United States. Although the wealthier Northeast states were successful in attracting German and British FDI during this period, poorer Southeastern states such as Florida (United Kingdom), South Carolina (Germany) and Georgia (both) were surprisingly successful as well (Figures VI.3 and VI.4).

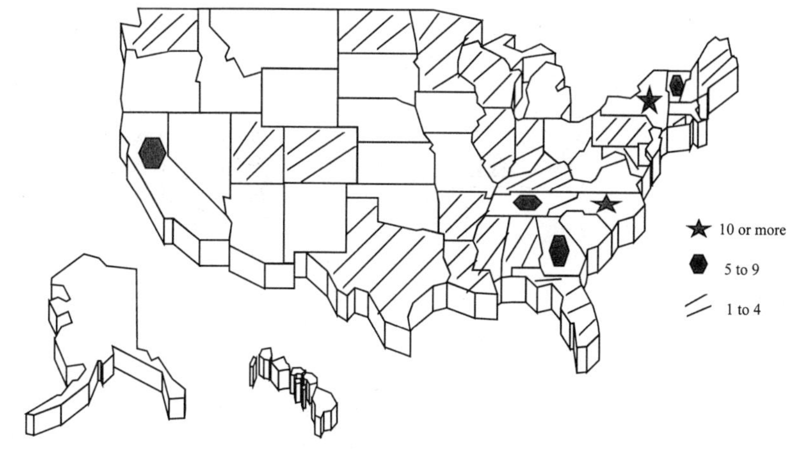

Source: Ondrich & Wasylenko (1993, p. 42).

Figure VI.5. Location of Canadian Manufacturing Plants in the United States, 1978 to 1987: 105 Total

It is not surprising that Canadian FDIs were concentrated in the Northeast, most notably New York state, but, once again, Southeastern states such as Tennessee, North Carolina and Georgia were relatively successful in competing for Canadian FDI capital (Figure VI.5). In short, in examining the pattern of manufacturing FDIs in the United States during the late 1970s and 1980s, it is evident that locational preferences may have been based in part on factors such as market size, income growth or resource wealth, but not exclusively.

In broadening the focus beyond manufacturing, what new observations, if any, can be made in reference to the location of new FDI in the United States in the 1990s? Table VI.2 and Figure VI.6 indicate the geographical distribution of new investments for 1993. Joining the large-market states of California, Texas, Michigan and New York, which traditionally have attracted disproportionately large shares of FDIs, were four states from the Southeast: North Carolina, Georgia, Florida and Virginia. Collectively, these eight states attracted approximately sixty-four percent of all new FDI in the United States for 1993 (Figure VI.6).

Seemingly, the pattern observed in the 1980s has continued in the 1990s. Some states have predictably attracted the attention and the capital of foreign multinationals based on the size of

Table VI.2. *State Location of Foreign Direct Investments in the United States, 1993*

States	#Of Class	#Of Cases/ Value Known	Value in Million $	States	#Of Class	#Of Cases/ Value Known	Value in Million $
Arkansas	-	-	-	Nebraska	-	-	-
Alaska	1	-	-	Nevada	6	3	93.1
Arizona	6	3	612.5	New Hampshire	-	-	-
Arkansas	1	1	98.0	New Jersey	13	4	57.0
California	77	30	2,776.7	New Mexico	-	-	-
Colorado	6	2	29.4	New York	70	25	7,022.2
Connecticut	10	4	1,89.3	North Carolina	16	9	420.5
Delaware	-	-	-	North Dakota	-	-	-
D. C.	6	1	240.0	Ohio	14	10	818.8
Florida	40	12	260.0	Oklahoma	5	3	688.2
Georgia	20	5	129.7	Oregon	7	5	1,267.7
Hawaii	2	-	-	Pennsylvania	13	7	340.3
Idaho	-	-	-	Puerto Rico	1	1	141.6
Illinois	13	7	425.6	Rhode Island	-	-	-
Indiana	5	3	79.5	South Carolina	10	4	115.8
Iowa	1	-	-	South Dakota	-	-	-
Kansas	7	3	354.6	Tennessee	9	3	149.5
Kentucky	6	4	323.8	Texas	33	21	4,104.0
Louisiana	7	5	1,466.8	Utah	2	2	880.4
Maine	0	-	-	Vermont	1	-	-
Maryland	8	2	145.0	Virginia	19	7	893.5
Massachusetts	13	7	2,020.0	Washington	7	3	118.0
Michigan	15	9	345.6	West Virginia	1	-	-
Minnesota	3	1	102.5	Wisconsin	2	2	212.3
Mississippi	2	1	80.0	Wyoming	1	1	120.5
Missouri	9	4	569.0	Unidentified	2	-	-
Montona	1	-	-	**TOTAL**	**491**	**214**	**28,961.4**

Source: U.S. Department of Commerce, Office of Trade.

their markets, the wealth of their resources and their agglomeration economies in general. On the other hand, other states, most notably in the Southeast, have been successful in attracting FDI despite limitations in the above. The question now arises, of course, to what extent have such states been successful, despite market and resource limitations, by compensating through the offering of artificial investment incentives?

THE NATURE OF STATE INVESTMENT INCENTIVE PROGRAMS

Intuitively, one would expect that market factors would play a dominant role in the locational decisions of foreign direct investors. Larger, wealthier states such as New York and California are able to offer large markets, world-class universities, skilled workers, efficient transportation systems and other attractive forms of social overhead capital. On the other hand, poorer states such as Georgia and Tennessee can offer relatively low wages and cheap, abundant land. Regardless of what natural advantage may emerge from a particular mix of resources or market characteristics, no state today, seeking new investments from abroad, can be totally oblivious to the need to sweeten deals with investment incentives.[4] Competition among states has reached such a level that states with unattractive market characteristics and limited resource endowments must seek to offset the same with artificial incentive packages, while states with market characteristics and resource endowments that translate into a comparative advantage must seek to protect the same through defensive investment incentive strategies.[5]

What was the genesis of this competition? The earliest state initiative designed to attract FDI was a promotional trip to Europe in 1959 by Luther Hodges, the governor of North Carolina (Kincaid 1984; Eisinger 1988). Word about the North Carolina initiative spread and, in the 1960s, visits to foreign countries became a staple gubernatorial activity. In the early 1960s, occasional visits gave way to regular visits and, towards the end of the decade, it became politically fashionable to establish permanent offices abroad in order to promote FDI on an ongoing basis (Eisinger 1988, p. 294).

Two major developments emerged in the 1970s. First, the U.S. government became involved in support of state promotional efforts. In 1971, the U.S. Department of Commerce, also interested in job promotion, provided funding for collective (multistate) investment missions overseas. This so-called "Invest in United States of America" program was managed by the National Association of State Development Agencies under Federal financial support (Eisinger 1988, p. 294).

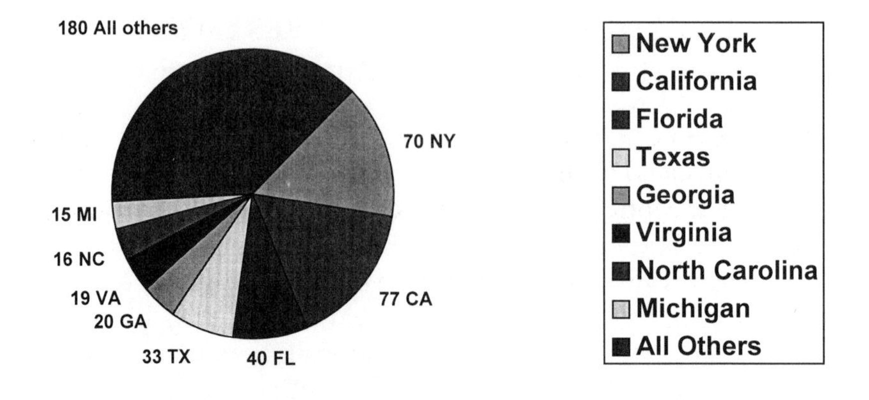

Source: U.S. Department of Commerce, Office of Trade and Economic Analysis.

Figure VI.6. Number of Foreign Investment by State 1993

Second, in the 1970s state efforts to attract FDIs became more formal. Earlier efforts by governors or their representatives to merely verbalize the virtues of their states' investment climates gave way to enabling legislature designed to lure new foreign business through state subsidies. In effect, rhetoric and moral suasion were replaced by official investment incentive programs.

According to a survey sponsored by the U.S. General Accounting Office, only ten states had committed funds to attract FDI prior to 1969. From 1969 to 1975, twenty-one states launched promotional efforts seeking foreign investment growth and from 1975 to 1978 fourteen more states had joined the group. As of 1979, the GAO reported that only three states had no foreign investment incentive program of any type (Kline 1982, p. 63).

Since the 1970s, seven major types of investment incentives have been employed by states in the competition for FDI. They are as follows:

- Financial Incentives: involving the use of a variety of financial tools available to assist investing companies on the cost side of the ledger. Instruments include loan subsidies, loan guarantees and industrial development bonds.
- Foreign Trade Zones: involving the establishment of trade areas into which foreign products may be brought with the payment of custom duties deferred until the goods leave the

zone entering the U.S. markets.

- Infrastructure Programs: including state government support of transportation, power, communications and other forms of social overhead capital. Housing development may be involved in this type of incentive package as well.
- Tax Incentives: offered in a variety of forms including tax credits, exemptions and rebates on both direct and individual tax levies. In cases where the expectations of foreign producers have not been met and leaving the state becomes a real threat, extended tax forgiveness or moratoriums have been employed to keep factories in state.
- Training Programs/Labor Incentives: providing recruitment assistance and training support in communities where skilled manpower is lacking and not capable of serving the needs of foreign producers. Support includes payment for the testing and screening of work force personnel as well as subsidies for radio, television and newspaper job advertisements. Contributions are sometimes made by states to local educational institutions to support curriculum efforts related to the needs of incoming foreign companies. These needs may be technical or non-technical.
- Natural Resource Programs: involve government assistance in providing foreign businesses with affordable industrial resources such as water, land and energy. In reference to land acquisition and development, properties needed by foreign producers may be contributed by the state, if state owned, or acquired under Eminent Domain, if privately owned. In areas where state environmental controls are more rigid than federal standards, such as pollution control, exemptive or special exclusions to state law are sometimes granted to the foreign producer.
- Enterprise Zones: providing specially designed areas in which foreign businesses are eligible for a package of investment incentives in addition to those available elsewhere in the state.[6]

In addition to the formal investment incentive programs, outlined above, that typically require enabling legislation by state governing bodies, there are informal arrangements, involving state influence or moral suasion, designed to entice foreign companies to invest locally. Potential foreign investors may receive promises of special treatment or special concessions if the need should arise. A good example is the offer made to Toyota by Martha Jayne Collins, the Governor of Kentucky in a 1985 letter. Specifically, she promised:

> In the event of any suit, protest or other complaint by residents, farmers or companies in the Georgetown area relating to the construction of the (Toyota) facility, the Commonwealth will mediate and use its best efforts to cause such suit, protest or other complaint to be settled as quickly as possible at minimum cost to Toyota (Newman and Rhee 1990, p. 65).

The 1980s witnessed an "explosion of activity" by states in competing for jobs through efforts to attract FDI (National Association of State Development Agencies 1986, p. 4). Rapid growth in industrial development marketing budgets occurred throughout the decade with increasing proportions going to overseas marketing efforts. However, not all budgeting outlays were earmarked for foreign investment promotion. Export promotion has also grown in importance and has become an important part of most states' industrial development programs (Eisinger 1988, pp. 290-306).

On the investment incentive side, some interesting changes have occurred over the past decade in the efforts of states to attract FDI. Promotional efforts have become more multidimensional and more high-tech. In addition to the more traditional approaches involving investment missions, trade shows and the opening of foreign offices, states are now employing electronic advertising and even video technology in delivering their message to potential foreign investors (National Association of State Development Agencies 1986, pp. 5-6).

Also, investment incentive programs are becoming more customized and company-specific, aimed at addressing the specific concerns that a foreign firm contemplating a move to the United States may have. If, for example, the foreign company being courted by the state has deep reservations about the adequacy of worker training in the state, budgetary flexibility typically permits a more concentrated state effort to allocate funds into human resource development. If high-tech worker skills are needed, state training programs will move in that direction.

Relating to customized incentive programs is the use of promotional packaging. States which aggressively seek to attract FDI essentially use the same tools. Thus, if investment incentives offered by several states appear to be equal to foreign companies, what becomes the next level of competition? The answer is, of course, the creative use of packaging; that is, combining different types of incentives and promotional actions in order to match the specific needs and preferences of potential foreign investors (Kline 1982, pp. 53-85).

If asked, state authorities will typically insist that investment incentives are available to local and foreign companies alike. It is true that most state subsidies are available to U.S. firms; however, it became clear in the late 1970s and 1980s that states were becoming particularly aggressive in seeking foreign investment capital through the use of customized incentive and promotional packaging approaches (Conway Data, Inc. 1991).

EFFECTIVENESS OF STATE INVESTMENT INCENTIVE PROGRAMS: EVIDENCE FROM THE LITERATURE

There is conflicting evidence in the literature concerning the relative importance of traditional state investment incentive programs in attracting FDI. On the positive side, a survey by the Japan Society of over 500 Japanese firms operating in the United States (Bob 1990) revealed that human resources incentives such as job training and tax incentives were factors in the FDI decision-making process. Survey respondents were asked to select the three most important factors in their parent corporation's choice of a U.S. investment location. The 585 participating companies responded as shown in Figure VI.7.

As indicated, human resource and tax incentives were factors but ranked well below "proximity to customers" and "quality of life" in terms of their relative importance in influencing the foreign investment decision making process.

The Japan Society also reveals that "greenfield" investments, which tend to be primary targets of state incentive programs because of the job creation potential, are particularly sensitive to human resource incentives. The study concludes that state-sponsored human resource development programs can be "powerful

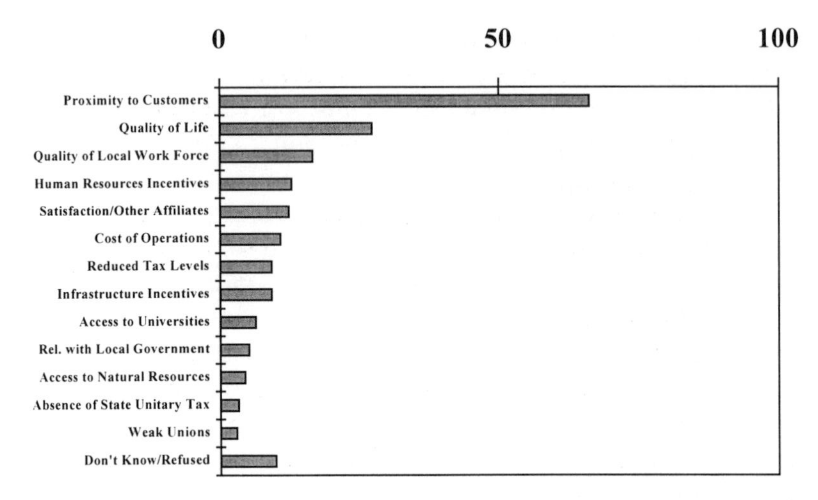

Source: Bob (1990, p. 43).

Figure VI.7. Survey Responses in Percentages

lures" in attracting Japanese greenfield investments (Bob 1990, p. 44).

A similar survey by Kim and Kim (1993) of approximately 100 Japanese multinationals lends support to the aforementioned Japan Society findings. Using a response scale from 5 (most important) to 1 (least important), state incentive packages were found to rank in the "middle" of factors that govern foreign investment decision making. The mean ratings for specific programs are in Table VI.3.

On the other hand, surveys and other empirical studies that include but extend beyond Japan, such as Glickman and Woodward (1989) and Sokoya and Tillery (1992) conclude that state incentive packages rank at the low end, rather than at the middle, of factors that induce foreign multinationals to come to the United States as producers. These studies suggest that it is more important for state and local authorities to nurture a positive political, social and economic environment that is conducive to business rather than "giving isolated preferential treatment to selected firms" (Sokoya and Tillery 1992, p. 78).

Tax policy tends to be at the core of state incentive packages and there have been several studies that have attempted to measure and weigh the impact of tax policy on foreign investment decision-making. Again, the results were mixed.

Table VI.3. *State and Local Government Incentives*

Rank	*Incentive*	*Mean Ratings*
1	Development of Infrastructure	2.84
2	Free Training of Workers	2.36
3	Tax Abatements	2.33
4	State-Sponsored Financing Arrangements	2.16
5	Free Land, Water and Electricity	2.12
6	Financial and Administrative Assistance	2.10

Source: Kim and Kim (1993, p. 68).

In the literature on business location, tax variables have been found to be statistically significant in a number of studies but not in others. Hartman (1984) concluded that tax rates have a significant effect on the inflow of FDI into the United States. His tests revealed that elasticities of FDI with respect to taxes, range from above unity to slightly above two, indicating sensitivity. On the other hand, Young (1988) computed smaller elasticities with respect to investments overall and found that FDI financed by transfers from parents were particularly insensitive to changes in tax rates. Root and Ahmed (1978) in an earlier study concluded that tax incentives fail systematically to attract FDI in manufacturing if matched by all competing countries but can be effective if the incentive tax package of a particular country is not matched.

Studies also reveal that FDI locational decision making seemed to be more sensitive to tax variables in the 1970s than in the 1980s. Indeed, many of the studies that uncovered foreign investor sensitivity to tax variables used 1970s data (Ondrich and Wasylenko 1993, p. 62). One plausible explanation was that significant tax level differentials existed in the 1970s because some states aggressively employed tax incentive programs and others did not. In the 1980s, most states entered the game and fiscal differentials among the states narrowed significantly.

Tax policy and tax rate differential matter, but tend to pale in comparison with these other factors governing the investment decision making. The most comprehensive examination of the link between tax policy and FDI locational decision making was conducted by Ondrich and Wasylenko (1993). The empirical results of this study revealed that fiscal factors rank well below market size and agglomeration economies in driving FDI. Since

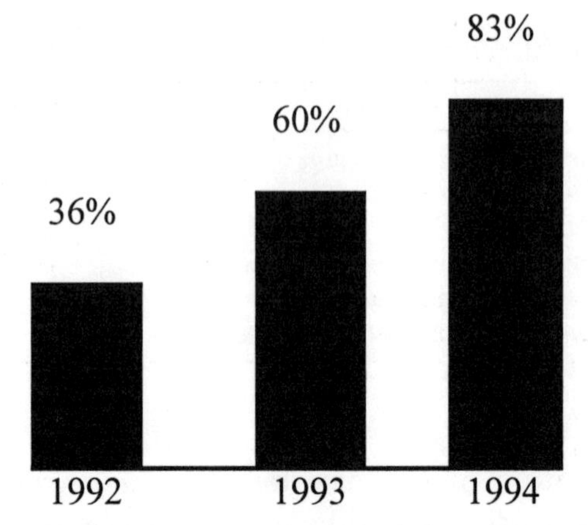

Source: *Site Selection: The Global Magazine of Business Strategy,* Conway Data, Inc. October, 1994, p. 858.

Figure VI.8. State Locational Incentives
[Percent Increases in New Incentive Packages]

states have almost no direct control over their agglomeration economies and the sizes of their markets, "few policy levers" are available to attract foreign investment, according to Ondrich and Wasylenko (p. 130). Nevertheless, the results of their testing do indicate that incentive programs offering corporate income tax relief may be effective in influencing the locational decisions of foreign multinationals (p. 134).

EFFECTIVENESS OF STATE INVESTMENT INCENTIVE PROGRAMS: NEW EVIDENCE

Have those states offering the most extensive investment incentive packages to foreign companies been successful in attracting the most FDI? The following analysis attempts to measure the effectiveness of incentive programs by comparing those states enjoying the heaviest infusion of foreign capital to those that are most aggressive in offering artificial investment inducements. Data sources include surveys from Conway Data and the Office of Community Planning and Development of the U.S. Department of Urban Development, supplemented by primary data gathered directly from state development agencies.

Table VI.4. *States Offering the Greatest Breadth of Coverage in Incentive Packaging (Out of a Total of Seven Program Categories)*

California	7	Louisiana	5
Kentucky	7	Michigan	5
Arkansas	6	Mississippi	5
Delware	6	New Jersey	5
Indiana	6	New York	5
Minnesota	6	Ohio	5
Missouri	6	Oregon	5
Virginia	6	Tennessee	5
Wisconsin	6		5

Source: Conway Data Surveys and U.S. Department of Urban Development, Office of Community Planning and Development, supplemented by primary data gathered directly from state development agencies.

Evidence clearly indicates that states, in general, have become more aggressive in seeking FDIs in the early 1990s. Figure VI.8 reveals the extent to which states have been accelerating their efforts in this regard.

In a Conway Data Survey of state development agencies conducted in June/July of 1994, it was revealed that the states of Alabama, Georgia, Indiana and Oklahoma offered the most new location incentives from 1992 to 1994. Interestingly, evidence offered earlier in the chapter (Figure VI.6 and Table VI.2) show that, with the exception of Georgia, these states do not rank among the largest, recent recipients of FDI capital.

However, it is necessary to examine existing levels of state investment incentives, in addition to recent changes in the same, in order to accurately assess effectiveness. The following is an effort to compare levels among states attempting to measure both the breadth and the depth of programs.

In the early 1990s, only two states (California and Kentucky) offered incentive packages in all of the seven traditional program areas outlined earlier in this chapter. Seven states, mostly from the Southeast and Midwest, participated in six out of seven, and eight states, mostly east of the Mississippi River, offered programs in five of the traditional areas (Table VI.4).

Once again, the attempt to link incentive programs to success in attracting FDI capital by measuring breadth of coverage in incentive packaging produces mixed results. Of the seventeen states cited in Table VI.4 as offering the greatest breadth in

Table VI.5. *States Offering the Most Extensive Packaging of Incentives Within Four Major Categories*

By State	***Enterprise Zone Programs*** *[Out of Total of 15]*	***Financial Assistance Programs*** *[Out of Total of 18]*	***Tax Incentives Programs*** *[Out of Total of 15]*	***Special Services Programs**** *[Out of Total of 18]*
Alabama	[8]			[15]
California	[11]			
Connecticut		[18]		
Delaware			[13]	[15]
Florida			[13]	
Illinois	[9]		[13]	
Indiana	[9]	[17]		
Iowa				[16]
Kansas	[8]			
Maine				[16]
Maryland	[8]	[18]	[13]	
Michigan	[9]			
Minnesota				[15]
Missouri			[13]	[16]
New York				[15]
North Dakota				[16]
Ohio		[17]		[15]
Oregon		[18]		[15]
Pennsylvania				[15]
Rhode Island			[13]	
South Dakota				[15]
Tennessee			[13]	
Texas	[8]		[13]	

Note: *Includes infrastructure, training and resource programs.

Source: *Site Selection: The Global Magazine of Business Statistics (October 1994, pp. 1150-1167)*

investment incentive packaging, only four states (California, New York, Virginia and Michigan) ranked among those states attracting the largest inflows of FDI capital in 1990-91 (Figures VI.6 and Table VI.2). Noticeably missing from Table VI.4 are such attractive foreign investment host states as Florida, Texas, Georgia and North Carolina.

Interestingly, one might surmise that market, resource and climatic factors in certain states in the Southeast may be sufficiently positive to attract FDI even in the absence of broad-based, investment incentive packaging.

Such a conclusion may be premature, however, because the evidence presented above is based on breadth of coverage in the offering of location incentives, but is silent on the question of depth of coverage. Are the so-called "hot" states successful in inducing foreign companies to locate within, not because of wide ranging incentive packages, but rather because of concentrated activity in certain key areas?

In Table VI.5, the attempt is made to measure the extensiveness of offerings in four major investment incentive areas, specifically, enterprise zone programs, financial assistance programs, tax incentive programs and special services programs.[7] According to Conway Data survey results, there were a total of eighteen types of financial assistance and special services incentive programs offered by the states in 1994.

Those states offering the most incentive packages in each program area are identified in Table VI.5. This is intended to measure depth of coverage as well as breadth. Interestingly, the data reveal that state allocations of incentive subsidiaries vary considerably. For example, Connecticut concentrates on financial assistance, offering every financial package available to foreign direct investors, but it does not rank high in the other categories. By way of contrast, Florida and Rhode Island rank at the top of those states offering broad packages of tax incentives (13 out of 15), but are not aggressive in other program areas.

In reference to both breadth and depth of coverage, only one state in 1994, Maryland, ranked high in three out of the four program areas (Table VI.5). Apparently, Maryland is atypical in its aggressive approach to investment incentive packaging by attempting to outgun its rival states through the offering of wide ranging enterprise zones, financial assistance and tax incentive programs. Maryland has had limited success in attracting FDI in the 1980s and 1990s and has not ranked with the leaders by any quantitative measure (Figures VI.5, VI.6 and Table VI.2).

Conway Data Survey data also reveals that, in 1994, eight states (Alabama, Delaware, Illinois, Indiana, Missouri, Ohio, Oregon and Texas) ranked high in their extensive incentive packaging in two out of the four aforementioned program areas (Table VI.5). Once again, the payoffs reaped by the states from their extensive resource commitments to incentive packaging are unclear. Of these eight states, only Texas and, to a lesser extent,

Illinois have ranked consistently high in the competition for FDI capital during the 1980s and 1990s.

On the other hand, states such as Georgia and Virginia have been remarkably successful in enticing foreign corporations to invest within their borders without the extensive packaging of investment incentives offered by several rival states. This is another piece of evidence in support of the notion that states' investment subsidies are not the dominant factors in the FDI decision making process.

THE HOT STATES IN THE 1990S: THE DOMESTIC CORPORATE VIEWPOINT

Since state investment incentive programs are used to attract the capital of domestic, as well as foreign corporations, two interesting questions arise: (1) Do the perceptions of U.S. corporations and foreign corporations on which states provide the best business climates coincide? (2) Does the effectiveness of incentive programs vary considerably or at all depending on whether the target is a domestic or foreign corporation?

Although the company-specific data necessary to fully address these questions are not presented here, Table VI.6 does shed some light on the issues. In a survey of predominantly U.S. corporations, the top business climates, state-by-state, are ranked based on an index of new business activity and corporate perceptions. Since 1993 data are included, it is possible to compare these rankings to the 1993 locational distribution of new foreign direct investment summarized in Figure VI.6 and Table VI.2.

Seemingly, based on this limited data, domestic and foreign corporations share the positive perception about the attractiveness of states in the Southeastern region as investment targets.[8] States such as Georgia, Virginia and North Carolina have attracted disproportionately large shares of both domestic and foreign capital in the early 1990s. Based on 1993 data, however, it would appear that the perception of domestic and foreign corporations differ in one significant area. Whereas both favor the Southeast and other sunbelt states, such as Texas, foreign corporations have been much more inclined of late to invest elsewhere, in the traditional large-market, heavily industrialized

Table VI.6. *Hot States*
Top 20 Business Climates (Based on New Business Activity and Corporate Business Climate Perceptions)

	Overall Rankings			
State	*1993*	*1994*	*1995*	*1996*
North Carolina	1	1	1	1
South Carolina	2	2	4	4
Georgia	3	6	7	8
Texas	4	4	3	3
Tennessee	5	9	6	6
Indiana	6	5	8	9
Kentucky	7	7	9	5
Virginia	8	11	11	7
Florida	9	10	5	12
Alabama	10	8	10	11
Missouri	11	14	—	—
Nevada	12	18	15	19
Ohio	13	3	2	2
Iowa	14	—	19	15
Louisiana	15	13	—	13
Arizona	16	17	13	9
Mississippi	17	19	17	—
Wisconsin	18	—	14	20
Illinois	19	12	—	—
California	20	—	16	17

Source: *Site Selection: The Global Magazine of Business Strategy*, (October, 1993, p. 1171, October, 1994, p. 959, October, 1995, p. 703, October, 1996, p.85).

states, particularly New York, Michigan and California (Figure VI.6).

In short, both domestic and foreign corporate capital has been moving south, but this regional movement seems to be more pronounced in the case of domestic corporations. Such an observation raises several questions worthy of future study. Are domestic corporations, in contrast to foreign companies, more interested in market growth or growth potential in a region as opposed to traditional market size? In moving south, does the domestic corporation place relatively more weight on cost factors, for example, low resource costs, than on revenue considerations? Is the southward movement of corporate capital grounded more in environmental or social concerns, rather than economic or financial conditions?[9]

Although these questions are worthy of serious study, they extend beyond the scope of this study. One related question is relevant, however. Do domestic corporations find state investment incentive programs to be more enticing than in the case of foreign corporations? The chapter has already established a weak link between state investment subsidies and FDI. Is the link any stronger in the case of the locational decision making of domestic corporate investors?

A re-examination of Tables VI.4 and VI.5 reveal that those states offering the greatest breadth and depth of coverage in incentive packages are not the "hot" states as perceived by domestic corporate investors. With the exceptions of Virginia, Kentucky and Texas, sunbelt states do not rank high in incentive packaging, but they have attracted the attention and the capital of U.S.-based corporations. In short, states' subsidies do not appear to be of primary importance in either the attraction of domestic or foreign direct investment.

CONCLUSIONS

To argue that state investment incentive programs are not the dominant factor in governing the locational decision of foreign direct investors is not to conclude that they are unimportant. The fact that all states employ some combination of incentive packaging indicates that state development agencies are able to project some form of investment return on such subsidies.

Survey results, reviewed in this chapter, do indicate that state investment incentives tend to rank at the low end, or in the middle, of factors cited as important in the response scales of foreign direct investors.[10] The new evidence presented in this chapter is consistent with such findings. Certain states, primarily in the sunbelt region, are viewed by foreign companies as superior investment locations because they possess the right combination of market, resource, environmental and social characteristics. These "hot" states apparently offer some investment incentives, probably for defensive purposes, but they do not feel compelled to offer the same extensive incentive packages as those states identified in the chapter as the most aggressive self-promoters.

To what extent can a state, faced with relatively unfavorable market or environmental conditions, offset such disadvantages by offering the most lucrative investment location subsidies? A satisfactory answer to the question must await further study, but it should be noted that those states that offer the most extensive incentive packaging do attract some foreign capital. They may not rank at the top of the list of those states that enjoy the heaviest infusions of FDI but neither do they rank at the bottom. Interestingly, a few rank near the top, but most can be found in the middle range.[11] To what extent these states are able to improve their relative positions as capital importers by offering incentive packages is not clear from the evidence.

What is clear from the data is the acceleration in the competition among states for FDI capital (Figure VI.8). This competition primarily takes the form of state subsidies of businesses and taxpayers have the right to know whether adequate investment returns are being generated by incentive programs.[12] The evidence would seem to indicate that taxpayers have reason to be concerned in this regard. Based on both the rhetoric and the resource commitments of most state development agencies, there seems to be a distorted view in state capitols about the relative importance of artificial investment incentives in the locational decision making of foreign MNCs. Depending on the particular state, its infrastructure, its resource mix and the company being courted, some investment returns on incentive packages may be substantial. However, given the magnitude of these programs and their widespread use, it is likely that the reality of investment returns in general may fall well below official expectations.

NOTES

1. There are several studies that examine factors that have motivated foreign firms to locate their productive activities in the United States. See Ajami and Ricks (1981), Caves and Mehra (1986), Chernotsky (1987), Graham and Krugman (1995), Hultman and McGee (1988), Martin (1991), Ondrich and Wasylenko (1993) and Sokoya and Tillery (1982).

2. Studies that examine linkages between FDI and market factors in host countries include Dunning (1988a), Franko (1971), Rugman (1981), Scaperlanda and Mauer (1969) and Williamson (1981).

3. The heavy infusion of FDI entering the United States during the 1970s, 1980s and 1990s has been well documented in the literature. See Graham and Krugman (1995) and Ondrich and Wasylenko (1993).

4. It should be noted that states design programs to attract foreign direct investment, not real estate or foreign portfolio investment. The reason is that the former is job-generating, particularly in the manufacturing sector. Thus, competing for FDI is really competing for jobs. For an overview of the nature of state-sponsored investment incentive programs, see OECD (1995, pp. 49-52).

5. See Conway Data, Inc. (1993, 1994 and 1995) for a comprehensive examination of the recent initiatives of the fifty state governments in competing for FDI capital.

6. The best source of information and data for state initiatives in these seven incentive areas are the October issues of *Site Selection and Industrial Development*, recently changed to *Site Selection: The Global Magazine of Business Strategy*, published by Conway Data, Inc.

7. Special services programs include state-sponsored job training and retraining, R&D promotions, public works projects and natural resource development.

8. Although the relative rankings of states preferred by domestic corporations (Table VI.6) shifted somewhat from 1993 to 1996, the most highly regarded business climates throughout this period were predominant from the Southeastern region. Only Texas, Indiana and Ohio were able to crack the top ten list of "hot" states from outside the Southeast during this time span and, within this group, only Ohio made a significant move to the top of the rankings.

9. Surveys of the attitudes of foreign direct investors have identified environmental factors, such as weather and social conditions, such as crime rate and population density, as being important in the locational, decision making process.

10. See Bob (1990), Kim and Kim (1993), Glickman and Woodward (1989) and Sokoya and Tillery (1992).

11. This conclusion is based on a visual analysis of Tables VI.1, VI.2, VI.4, VI.5, VI.6 and Figures VI.1, VI.5.

12. There is some evidence that the sensitivity of foreign direct investors to state-sponsored investment incentives vary from industry to industry. Yoffie (1993) observed that FDI in the United States in the semiconductor industry has tended to cluster in those states offering the most extensive incentive packages. Based on this observation, he speculated that foreign multinationals in capital intensive industries tend to be more receptive to incentive packages than others. Confirmation of this linkage obviously must await other industry studies of this type.

Chapter VII

Conclusions

It should not be surprising that recent trends in foreign direct investment have captured the full attention of a diverse group of casual observers, professional organizations and vested interests that track important developments in the global economy. These include the commercial press, scholarly associations, professional think tanks as well as national and international advisory services and policy making groups.

International organizations, such as the OECD, GATT and the WTO, which historically focused on foreign trade related issues and problems, have reallocated much more research time and effort over the past two decades in examining the causes and consequences of the dramatic, global growth of foreign direct investment. Government officials and politicians, who historically debated the social benefits and costs of opening national borders to the international movement of merchandise, have become much more inclined of late to broaden the debate to include the international flow of capital, including FDI. Scholarly journals in the field of international economics, once inclined to favor research products that shed light on the processes and dynamics of trade among nations, have become much more receptive to scholarly efforts that examine the phenomenon of FDI.

The reasons for the significant growth of interest in FDI can be expressed in both quantitative and qualitative terms. In reference to the former, a recent report of the World Trade Organization (1996), revealed that the annual global flow of FDI between 1985 and 1995 increased astronomically from around $60 billion to approximately $315 billion. Moreover, UNCTAD data suggest that, by 1995, the sales of the foreign affiliates of multinational corporations has grown globally to exceed the value of world trade in merchandise and services. Finally, there is growing evidence that foreign trade has become closely linked to FDI and multinational firm activity. In 1995, approximately one-third of total world trade involved intra-firm trade among MNCs, and MNC exports to non-affiliates accounted for another third of aggregate trade (United Nations Centre on Trade and Development 1995). Clearly the relationship between FDI and world trade has become complementary, not substitutional.

On the qualitative side, there is growing evidence that FDI is bestowing benefits on host countries, investing countries and MNCs per se in various ways. FDI is viewed as an efficient way to allocate the world's scarce resources to promote both productivity and economic growth, to transfer technology horizontally and vertically across national boundaries, to transfer organizational management and marketing skills and to propel the process of globalization that is contributing to the ongoing integration of the world economy (World Trade Organization 1996, p. 4).

Of course, the growing interest in FDI that has emerged over the past two decades has not been based on real or imagined benefits alone. Concern has been expressed and continues to be expressed in both the commercial and scholarly literature about the potential social and economic costs of FDI, including adverse employment, real wage rate, trade balance and national security effects.[1]

The dramatic growth of FDI globally during the 1980s and 1990s has stimulated two major streams of research, the first focusing on the motivating factors and root causes of the surge and the second on the effects and consequences. It is the purpose of this book to contribute to the former.

THE EVOLUTION OF FDI THEORY

In the 1950s and 1960s, FDI theory developed primarily in the attempt to identify the factors responsible for the heavy outflow of FDI capital from the United States during this early period. Subsequently, the maturing of FDI theory coincided with the surge in inward FDIUS which began in the 1970s and accelerated throughout the 1980s.[2]

The deceleration of both inward FDIUS and outward USFDI in the early 1990s had little impact on scholarly interest in theory with the possible exception of reinforcing the important role of macroeconomic conditions and structural change in the global economy as governing factors. The renewed surge of FDIUS in the mid-1990s, as documented in Chapter I, has lent support to the notion that developments in the 1980s were not an aberration and that FDI is an ongoing phenomenon worthy of the full attention of scholars specializing in the study of the global economy.

In the maturation of FDI theory from the 1950s to the 1990s, significant transformations have taken place. Early theory was limited in the sense that it focused on the statics of why firms go overseas as foreign direct investors, that is, explaining entry, rather than on the dynamics of MNC strategies of global, corporate growth. In the 1950s and 1960s, FDI theories were primarily formulated by economists who borrowed heavily from traditional microeconomic analysis. FDI motivations were attributed to market imperfections which created opportunities for firms to go overseas as producers and convert firm-specific assets into comparative advantage in competition with local firms. In this early theoretical work, these "assets" were identified initially as tangibles, such as a superior product or production process. Later, attention turned more to intangible assets such as superior managerial and organization skills or superior marketing know-how.

Predictably, FDI theory moved off in several directions as it matured in the 1970s, including a focus on the intra-firm rivalry and oligopolistic behavior of MNCs. However, the mainstream research of this period, in examining the motives of MNCs, began to look more inwardly, uncovering ways that firms can exploit intangible assets profitably by internalizing international operations. Theories were designed to demonstrate how firms are able to justify the added costs of managing large, geographically dispersed organizations by effectively capturing comparative advantage through the internalization of global transactions, particularly in the market for knowledge and technology.

FDI theory, which was dominated in the 1950s and 1960s by work that used traditional economic analysis as a springboard, became more multidimensional and multidisciplinary in the

1970s and 1980s. The development of international business management as a subject of scholarly interest led to a cross fertilization of ideas descending on the study of FDI from several functional areas of management science. Thus, the paradigms formulated during this period transcended the normal boundaries of economics, extending to and beyond the perspective provided by international finance, international marketing and organizational theory. While some researchers in the 1980s and 1990s have continued to draw from the traditional thinking of economists and the more recent thinking of management scientists in the formulation of FDI theory, others have succeeded in broadening the boundaries of scholarly inquiry even more extensively by providing more of a sociocultural, political framework.

Perhaps most importantly, the focal point of FDI theory formulated over the past decade has moved beyond the issue of entry. Less attention is paid to the factors governing the initial entry of the MNC into the host country and more attention is directed at what happens beyond that point. What are the strategic factors that govern the nature, pace and direction of MNC investment as the organization expands its operations globally? Modern FDI theory is designed to capture the dynamics of the MNC investments, reinvestments and growth.

FDIUS MOTIVATION: THE EVIDENCE BASE

The evolution of FDI theory since the 1960s has predictably been accompanied by ongoing efforts by scholars to provide empirical verification. The techniques employed have included the full range of possibilities including surveys, case studies, time series analysis and cross sectional analysis.[3]

Surveys and case studies have revealed that the factors motivating FDIUS over time have been multidimensional and multifaceted, changing from time to time, from company to company and from country to country. This diversity of motivational factors has, on one hand, presented opportunities for a multitude of empirical study designs, approaching the topic from different directions and with different focal points. On the other hand, the complexities of motivating factors have created difficulties in the area of statistical testing, including the problem of finding appropriate proxies for such intangible, but relevant, factors as the stability of the U.S. political system and the cultural characteristics of U.S. society.

Of course, some motivating factors are both concrete and visible and, accordingly, have been universally identified by empirical studies. The "size" of the U.S. market emerged from most early empirical work in the 1960s and 1970s as a powerful magnet drawing FDI to U.S. shores. The results of these early studies raised a puzzling question, however. If the large U.S. market was such a dominant factor in attracting FDIUS during the 1960s and 1970s, why did comparably large U.S. markets during earlier decades fail to do the same?

Evidence in the literature provides a clear answer to the above. The large U.S. market has always whetted the appetite of foreign MNCs, as has the attraction of having a production presence in the United States and the ability to use that presence in obtaining technical and managerial knowledge from contact with local U.S. firms. However, until the 1960s and 1970s, the perception of U.S. corporate superiority served as an effective entry barrier, discouraging FDIUS. The breakthrough came during the 1960s when European firms began to compete successfully with U.S. multinationals in Europe, exploiting firm-specific advantages, and realized that those same "assets" could also translate into a comparative advantage in the U.S. market. In short, studies revealed that FDI, particularly from Europe, was both pushed and pulled to the United States in the 1960s and 1970s. The vast size of the U.S. market and the opportunity to learn from a production presence in the United States were magnets pulling FDI to U.S. shores. However, it was the emergence of competitive advantage in the form of firm-specific assets, such as superior product or production technology, that pushed foreign firms to seek to exploit such advantages in the Western Hemisphere.

Predictably, some of the empirical studies conducted during the 1980s and 1990s have been designed to retest earlier formulated theories. Results have been mixed, but more recent evidence does generally lend further support to the notion that MNCs are motivated to enter foreign markets as producers if they possess firm-specific resources or assets that are capable of being translated into competitive advantage.

As indicated above, such advantages were initially thought to be based on tangibles, such as differentiated (i.e., superior) products. However, with the evolution of empirical studies over the past three decades, it has become increasingly more evident that competitive advantage can be based on several forms of intangible assets such as intellectual property, organizational skills or marketing know-how.

More recent empirical work has also been successful in identifying the specific competitive advantages that MNCs gain from internalizing the exploitation of intangible assets in overseas operations. FDI has been shown to flow from ownership-specific advantages, such as patents, exploited in conjunction with location-specific factors such as low material or labor costs. This evidence reaffirms that FDI is both "pushed and pulled" to foreign shores.

Since early FDI theory tended to be microeconomic rooted, early empirical studies (1960s and 1970s) focused very little attention on macroeconomic determinants or political/cultural factors motivating FDI. More recent studies (1980s, 1990s) have shifted attention to the latter with the evolution of FDI theory in that direction. Although case studies and surveys have documented the importance of sociopolitical and cultural conditions in the host countries to the investment decision making of foreign MNCs, confirmation by more scientific studies has been impeded by the difficulty of finding appropriate proxies for such intangible motivational factors. On the other hand, the importance of macroeconomic conditions and imbalances, as factors governing the flow of FDI, has become well documented, particularly in the case of Japan.

THE SPECIAL CASE OF JAPAN

In examining the characteristics of FDI in the United States in recent decades by country of origin, it is clear that Japanese investments have different quantitative and qualitative dimensions. Swings in Japanese FDIUS have been more pronounced than in the case of any other investing country, increasing dramatically in the 1970s and 1980s and retreating dramatically in the early 1990s prior to a sharp, recent recovery.[4] Also, Japanese MNCs in recent decades have demonstrated unusual skill in exploiting ownership-specific advantages, such as superiority in certain types of product and process technology, in the U.S. market.

In a sense, Japanese MNCs have been motivated by the same "pull" factors as in the case of MNCs from other countries. The difference has been in the "degree" not in the "kind." All countries have been attracted by the size and positive characteristics of the U.S. market and have been drawn to the U.S. shores by the quest to obtain new knowledge. Furthermore, FDIUS from all investing countries has been artificially induced by the threat and the reality of U.S. trade protectionism. What has set Japanese FDIUS apart from others has been the extreme sensitivity by which Japanese MNCs react to these "pull" factors.

Evidence indicates that macroeconomic conditions in Japan, in relation to those in the United States, have been very instrumental in "pushing" Japanese investments to U.S. shores. Since the macroeconomic imbalance between the United States and Japan in recent decades has been greater than between the United States and any other investing country, the fact that Japanese FDIUS has been more volatile than that of other countries should not be surprising.

Macroeconomic effects include the savings/investment imbalance in Japan during the 1980s and early 1990s at which time private and public savings grew in excess of business and government investments domestically. This pushed surplus funds overseas, including FDI. Since the opposite savings/investment imbalance existed in the United States, a disproportionate share of Japanese capital moved to the United States. The creditor/debtor relationship between the two countries can be clearly linked to significant national differences in the relative propensities to save and invest both privately and publicly.

Persistently large Japanese current account surpluses during the late 1970s, 1980s and 1990s, in the face of comparably large U.S. deficits, also have served as a closely related macroeconomic condition pushing Japanese FDI to U.S. shores. The Japanese government's reaction to the growth of external payment surpluses reinforced this international investment effect. Faced with the external "image" problem of having growing surpluses, exchange reserve accumulation and exchange rate distortion, the Japanese government released pressure over time by inducing Japanese MNCs to move their capital and operations overseas through loan subsidies and other incentive programs.

Throughout the time period of this study there has been a strategic element in Japanese FDIUS. Early investments (1950s and 1960s) were, in part, strategic responses to natural resource and material shortfalls in Japan. Later (1970s and 1980s), Japanese FDIUS became, again in part, a defensive response to the threat of a U.S. governmental protectionist crack-down on Japanese export surpluses. Throughout the past three decades, Japanese multinationals, organized in keiretsu groups, have strategically sought to capture economies of scale and scope externally as they have domestically throughout their recent history.

On the dark side, it has even been suggested that the Japanese industry/government strategy involves exploiting the relatively open U.S. FDI environment while maintaining restrictions on U.S. corporate activity in Japan, thereby giving Japanese MNCs a competitive edge on their U.S. corporate counterparts in the global economy.[5] Whether such asymmetric FDI policy is a carefully conceived strategic design or not, it is clear that public policy initiatives do shape and govern FDI flows to some extent. To what extent has the U.S. government been proactive in this regard?

FDIUS AND U.S. GOVERNMENT POLICY

Historically, the U.S. Government has not been proactive in using its fiscal or monetary powers to promote the inflow of FDI. Investment incentive programs designed to attract the attention of foreign MNCs exist in the United States, but they are sponsored by state governments, not the Federal government.[6] Prior to the 1980s, the U.S. government was quite consistent in treating foreign firms seeking entry or already operating on U.S. soil with openness and neutrality. Historically, this has involved the extension of "national treatment," that is, the right of foreign firms to establish subsidiaries in the United States and the right, once established, to receive the same regulatory treatment as local firms.

Although officially U.S. policy remains supportive of free international capital flows, developments in the 1980s and 1990s have caused some erosion in the U.S. government's commitment to openness and neutrality in FDI policy. Astronomical U.S. current account deficits in the 1980s, offset in part by heavy infusions of inward FDI, led U.S. politicians to agonize over the absence of a "level playing field" and decry the loss of "economic sovereignty." Structural changes in the 1990s have altered U.S. payments imbalances with the rest of the world and have disrupted the trend of accelerating inward FDIUS, but the concern over foreign acquisition of U.S. property assets has continued in Washington, DC.

This concern has translated into a number of proposed laws aimed at restricting the entry and/or growth of foreign firms in the United States. Most proposed pieces of legislation have failed to be enacted because of consistent and effective opposition from the Reagan, Bush and Clinton administrations. However, there have been exceptions, such as the Exon-Florio amendment to the Omnibus Trade and Competitiveness Act of 1988, which authorizes the blockage of inward FDI based on real or imagined threat to U.S. national security.

Although no administrative order or legislative act in Washington over the past two decades can be clearly interpreted as a clean break with the traditional U.S. commitment to FDI openness and neutrality, increased official sensitivity to reciprocity issues and to the competitive opportunities of U.S. multinationals abroad has become quite visible. Non-discriminatory U.S. treatment of foreign direct investors, which historically was unconditional, has become increasingly more conditional.

Internationally, the U.S. government continues to negotiate bilaterally and multilaterally with the expressed purpose of promoting freer international capital flows and a more open FDI climate. Unfortunately, the recent shift in the focus of U.S. government policy and practice to issues relating to national security, reciprocity, and "level playing fields" has tarnished the country's image as leader in the global effort to liberate foreign investment flows and liberalize foreign investment policies.

STATE GOVERNMENT POLICY AND FDI INCENTIVES

The U.S. Federal government may not compete for FDI though the offering of artificial investment incentives, but state governments do, and the competition among states has become increasingly more aggressive over the past three decades.[7]

At the state level, programs designed to attract FDI include: (1) financial incentives, such as loan guarantees; (2) the establishment of foreign trade zones with custom duty deferrals; (3) infrastructure programs, in which state governments provide social overhead capital; (4) tax credits, exemptions and rebates; (5) recruitment assistance and training support for workers; (6) natural resources programs involving government support in providing foreign businesses with affordable industrial resources; and (7) the establishment of enterprise zones, in which the locational decisions of foreign MNCs are rewarded with special subsidies.

Empirical evidence, including the primary data presented and analyzed in this study, seems to indicate that state investment incentive packages have some modest impact on FDI decision making, but their relative importance is low compared to market incentives in the private sector of the economy. In surveys, state investment incentives consistently rank at the low end or low/middle range when respondents are asked to rank all motivational factors in a scale of importance.

It is true that all "hot" states, that is, those that have attracted the most FDI in recent years, offer artificial incentives. However, these tend not to be the most aggressive states, offering the most extensive and lucrative incentive packages to foreign companies. Aggressive states tend to attract FDI, but they do not rank at the top of the list of hot states. Neither do they rank at the bottom.

The effectiveness of state incentive programs tends to vary from company to company and from country to country. For example, tax holidays may be attractive to foreign companies from countries that do not permit write-offs of taxes paid overseas but most unattractive to companies that receive generous credits on local returns for foreign tax liabilities. On balance, market factors seem to be more important than artificial incentives in governing the locational decision making of foreign direct investors. The "hot" states in the United States seem to be those that offer some incentives, but more importantly, have favorable market characteristics and social and cultural environments that promote a good quality of life.

THE FOCUS AND DIRECTION OF FUTURE RESEARCH, WHERE IS IT HEADING, AND WHAT NEEDS TO BE DONE

The resurgence of FDIUS in the mid 1990s, following a slowdown in the early part of the decade, has once again sparked interest in the causes, consequences and implications of the phenomenon. Some of the old questions and issues raised in this study continue to be relevant with the approach of the 21st century. To what extent and in what ways are factors currently motivating FDIUS similar or dissimilar to the past? How will this evolve in the future? What will be the relative importance of market factors vis à vis public policy incentives in attracting foreign producers to U.S. shores into the next century?[8] In reference to motivational factors in the future, what will be the relative importance of macroeconomic conditions in the United States and global economies, compared to microeconomics and managerial concerns and considerations? In the motivational mix, what role will be played by the socio-political and cultural climate of the United States compared to purely economic factors?

Although the old questions remain important, new questions are now being raised which, in the future, will hopefully add fresh insight and a better understanding of what is driving and will continue to drive FDIUS. It has been revealed in this study that FDIUS motivation can vary from country to country. For example, Japanese motivation to invest in the United States was shown to be atypical in recent decades because of the huge trade imbalance and contrasting macroeconomic conditions between the two countries. Recent research has focused some attention on factors motivating FDIUS by "country of origin of investment" (Grosse and Trevino 1996), but more needs to be done.

The "lion's share" of FDIUS continues to originate from other industrialized countries, justifying a focus on such investment relationships. Although FDI inflows from third-world countries and from countries in transition from central planning to market-based capitalism are a small percentage of the total FDIUS, the trend is positive, and recent growth is worthy of serious study.[9]

Traditional foreign investment relationships are changing in the global economy and will continue to change as economic and political reform continues in Eastern Europe and China and as the industrial powers of East Asia struggle to overcome financial instability. In addition to the increasing capabilities of several emerging and developing countries to move production facilities overseas, including to the United States, they are also becoming more attractive as targets of FDI and the resulting investment diversion effect will clearly impact the United States (Banks 1996).

The net effect of all this is not clear, however, because political reform and economic change in the world economy are altering comparative advantage among counties and companies and are changing the way in which MNCs organize and locate their production facilities overseas. In the literature, this is already the call to reappraise global corporate strategies and to revise FDI theory accordingly (Dunning 1995). With structural change in the economic and political underpinning of the global economy, a new trade theory has emerged in the economic literature. A new FDI theory can not be far behind (Markusen 1995).

In the past, international trade problems and issues received more attention than international investment flows from national and international policymakers. More recently (1980s and 1990s), attention has shifted to the latter in part because of the rapid growth of international investments, including FDI, and in part because of evidence revealing that the relationship between trade and investment globally is more complementary than substitutional (World Trade Organization 1996, pp. 18-22).

In tracing the root causes of FDIUS in the future, more attention should be paid to trade linkages both on the export and import side. To what extent and in what sectors of the economy is there a complementary (or any other) relationship between FDIUS on one hand and U.S. exports and imports on the other? How might changes in U.S. trade policy affect inward FDI? Can trade policy be used to govern inward FDI to any meaningful extent?

On the microeconomic level, more research needs to be focused on small and medium sized transnational corporations. Are smaller firms, contemplating moving their operations overseas, motivated by the same factors or conditions that influence larger firms? Are there industry-to-industry differences in foreign investment motivation based on firm size differentials within such industries? Historically, research of this type has been limited by the paucity of disaggregated data on FDI and the tendency to select large firms for case studies, again based on data availability. However, since firm-specific data recently have become more available, for example, from the UNCTAD Program on Transnational Corporations, studies on the FDI activities of small and medium-size firms are beginning to emerge (Fujita 1995). This is a beginning, but much more data and data analysis are needed.

Recent research on FDI has moved in several other new directions. Traditionally, theoretical work has offered rich explanations of the "why," "where," and "who" of FDI. Borrowing from the economics of uncertainty, new studies are beginning to examine the timing of FDI, focusing more on the question of "when" (Rivoli and Salorio 1996).

The effectiveness of artificial incentive programs offered by state governments in the United States, in the attempt to attract FDI, is unclear from the evidence presented in this study. This is generally true of other studies raising the same question in reference to governmental incentive programs of other nations. In fact, empirical studies have arrived at contradictory results, particularly in reference to the effectiveness of tax-related programs.[10] More work needs to be done not only in evaluating the effects of these programs on the locational decision making of foreign MNCs, but also in measuring the relative importance of policy and non-policy variables in the decision-making process. The degree of difficulty of seeking clarification in this regard through statistical analysis has been addressed recently in the literature (Loree and Guisinger 1995).

The ultimate attraction for all investors, including MNCs, is the rate of return on investment (ROI). Of course, for the firm weighing the merits and demerits of a foreign direct investment, the governing factor is the expected ROI in relation to opportunity costs. More research needs to be done on: (1) how MNCs project expected returns, (2) how ROI on current overseas operations affects reinvestment rates in the same host countries or new investments in different target areas and (3) how government policies in both the investing and the host country affect the ROI of foreign direct investments. A few recent studies (Pechter 1994; Lessard 1995) have shed some light on these and related issues, but more comprehensive research, using broader data bases, are clearly needed.

Finally, the multinational corporation and its global strategies have become the object of extensive multidisciplinary study over the past two decades. Growing interest in the MNC from scholars, examining its behavior from several disciplinary directions, has led to the maturation of FDI theory. Despite progress, further interdisciplinary integration is needed to permit the development of a truly general theory of FDI. Explaining the factors governing the entry of the MNCs into foreign markets is relatively easy. Explaining the strategic growth of MNCs, including how firms achieve synergy through international organizational linkages or develop competitive advantages globally through strategic alliances, is clearly more difficult.

The phenomenal growth of foreign direct investment worldwide in recent decades, including FDIUS, has raised questions, issues and tensions. The public policies of national governments have not evolved as rapidly, nor in some instances even in the same direction, as corporate strategies.[11] MNCs may wish to operate freely and efficiently across national borders, but the world economy currently is imperfectly integrated and nation states through their policies and regulations continue primarily to pursue national, not global, objectives. Accordingly, the scholarly examination of the MNC and its global strategies should not be conducted in isolation, but rather within the context of the international legal and regulatory framework, including the laws and regulations of both investing and host country governments.

The future research agenda is lengthy and opportunities to generate new ideas and make new contributions to the understanding of the process of FDI will likely multiply. All evidence points to the continued globalization of business and the ongoing interest on the part of scholars, corporate managers, government policy makers and others in understanding the process.

NOTES

1. For a comprehensive analysis of these effects, see Graham and Krugman (1995, pp. 57-120).

2. The evolution of FDI theory over the time period of this study is examined in detail in Chapter II. Of the references cited, several contain good reviews

of the literature on FDI theory; these include Casson (1986), Graham and Krugman (1995), McCulloch (1993) and Ondrich and Wasylenko (1993).

3. Chapter III focuses on efforts to provide empirical verification for FDI theory in the identification of factors motivating FDIUS. For interesting overviews, see Ajami and Ricks (1981), Sokoya and Tillery (1992) and United Nations Centre on Transnational Corporations (1992).

4. The central theme of Chapter IV is that Japanese FDIUS is a special case both quantitatively and qualitatively. For comprehensive summaries of factors motivating Japanese FDIUS, see Chernotsky (1987) and Kim and Kim (1993). For some interesting data and data analysis, contained in a study sponsored by the Japan Society, see Bob (1990).

5. The strategic dimensions of Japanese FDIUS are examined in Graham and Krugman (1995, pp. 66-67).

6. Although the Federal government of the United States, unlike other national governments, does not seek to compete for FDI capital by offering investment incentive packages, U.S. policy does influence FDIUS in both positive and negative ways. The specific U.S. policies and practices that govern or influence inward FDI are the focus points of Chapter V.

7. The best source for tracking trends and changes in the investment incentive programs of state governments in the United States is a monthly magazine published by Conway Data, Inc. titled *Site Selection: The Global Magazine of Business Strategy* (formerly called *Site Selection and Industrial Development*). Investment incentive data are presented in the October issue of the magazine. In Chapter VI, the analysis of state incentive programs is based on secondary data (from Site Selection) and on primary data (gathered directly from state economic development agencies).

8. For an interesting recent study of the effects of policy and non-policy variables on the location of USFDI abroad, using data from the 1970s and 1980s, see Loree and Guisinger (1995).

9. For a discussion of trends in FDIUS, see Borghese (1993). Relevant data on such trends appear annually in the July or August editions of the Survey of Current Business.

10. A summary review and analysis of the findings of studies that examine the impact of taxation on FDI can be found in this book in Chapter IV, V and VI.

11. The tensions and inconsistencies that exist in the corporate/government interface, with reference to FDI, are examined cogently in Graham (1996, Chap. III). Recent efforts to resolve these differences are discussed later in the book (Chapters IV-VI).

REFERENCES

Ajami, R. and D. Ricks. 1981. "Motives of Non-American Firms Investing in the U.S." *Journal of International Business Studies* 12: 25-34.

Aliber, R. 1970. "A Theory of Foreign Direct Investment." In *The International Corporation*, edited by C. P. Kinleberger. Cambridge, MA: MIT Press.

______. 1983. "Money, Multinationals and Sovereigns." In *The Multinational Corporation in the 1980s*, edited by C. P. Kinleberger and D. B. Audresch. Cambridge, MA: MIT Press.

American Embassy, Tokyo. 1994. *Japanese Foreign Direct Investment: A Report on the 1993 Japan Export Import Bank Survey*, March 17, pp. 1-3.

Anderson, A. and K. Noguchi. 1995. "An Analysis of the Intra-Firm Sales Activities of Japanese Multinational Enterprises in the United States: 1977 to 1989." *Asian Pacific Journal of Management* 12: 69-89.

Anderson, G. 1988. "Three Common Misperceptions About Foreign Direct Investment." *Federal Reserve Bank of Cleveland* 1-4.

Andersson, T. 1992. "The Role of Japanese Foreign Direct Investment in the 1990s." Working Paper of the Industrial Institute for Economic and Social Research. Stockholm, Sweden.

Ando, A. and A. Auerbach. 1988. "The Cost of Capital in the United States and Japan: A Comparison." *Journal of the Japanese and International Economies* 2(2): 134-158.

APEC Ministers' Joint Statement. 1993. *Asia-Pacific Economic Co-operation Ministerial Meeting*, November 17-19, Seattle, Washington.

Awanohara, S. 1992. "Lost Incentive: Japanese Investors Turn Away from Uncle Sam." *Far Eastern Economic Review* 10: 58.

Bailey, M. and G. Tavlas. 1992. "Exchange Rate Variability and Direct Investment." *The Annals of the American Academy of Political & Social Science* 106-116.

Banerji, K. and R. Sambharya. 1996. "Vertical Keiretsu and International Market Entry: The Case of the Japanese Automobile Ancillary Industry." *Journal of International Business Studies* 27: 89-113.

Banks, H. 1996. "Who Gets the Foreign Direct Investment?" *Forbes* 155 (April 10): 41.

Bartik, T. 1991. *Who Benefits from State and Local Economic Development Policies?* Kalamazoo, MI: Upjohn Institute for Employment Research.

Bartlett, C. and S. Ghoshal. 1989. *Managing Across Borders*. Boston, MA: Harvard Business School Press.

Beechler, S. and J. Yang. 1994. "The Transfer of Japanese-Style Management to American Subsidiaries: Contingencies, Constraints and Competencies." *Journal of International Business Studies* 25: 467-491.

Bergsten C., T. Horst and T. Moran. 1978. *American Multinationals and American Interests*. Washington, DC: Brookings Institution.

Berzirganian, S. 1991. "U.S. Affiliates of Foreign Companies: Operations in 1989." *Survey of Current Business* 17: 72-92.

Bhagwati, J., E. Dinopoulos and K. Wong. 1992. "Quid Pro Quo Foreign Investment." *AEA Papers and Proceedings* 82: 186-190.

Birkinshaw, J. and A. Morrison. 1995. "Configurations of Strategy and Structure in Subsidiaries of Multinational Corporations." *Journal of International Business Studies* 26: 729-753.

Board of Governors of the Federal Reserve System. 1996a. *Federal Reserve Bulletin* 82: 701-716.

______. 1996b. *Federal Reserve Bulletin* 82: 210-213.

______. 1995. *Federal Reserve Bulletin* 81: 832-837.

______. 1994a. *Federal Reserve Bulletin* 80: 782-785.

______. 1994b. *Federal Reserve Bulletin* 80: 584-588.

Bob, D. 1990. *Japanese Companies in American Communities*. New York: Japan Society.

Boddewyn, J. 1983. "Foreign Divestment Theory: Is It the Reverse of FDI Theory?" *Weltwirtschaftliches Archiv* 119(2): 345-55.

______. 1988. "Political Aspects of MNE Theory." *Journal of International Business Studies* 19: 1-31.

Boner, B. and A. Neef. 1977. "Productivity and Unit Labor Costs in 12 Industrial Countries." *Monthly Labor Review* C: 16.

Bonito, G. and G. Gripsrid. 1992. "The Expansion of Foreign Direct Investment: Discrete Rational Location Choices or a Cultural Learning Process?" *Journal of International Studies* 23: 461-476.

Borghese, K. 1993. "Developments and Trends in Foreign Direct Investment in the United States." In *Economics and Statistics Administration, Office of Chief Economist, Foreign Direct Investment in the U.S.: An Update*. Washington, DC: U.S. Government Printing Office.

Boskin, M. 1987. *Reagan and the Economy: The Success, Failures and Unfinished Agenda*. San Francisco: ICS Press.

Bradley, G. and J. Lewis. 1979. "The U.S. Market: A Good Catch for Investors." *Vision: The European Business Magazine* 102: 43-53.

Buckley, P. and M. Casson. 1976. *The Future of the Multinational Enterprise*. London: Macmillan.

Buckley, P. and J. Dunning. 1976. "The Industrial Structure of U.S. Direct Investment in the UK." *Journal of International Business Studies* 7: 1-16.

Buckley, P. and R. Pearce. 1979. "Overseas Production & Exporting By the World's Largest Enterprises: A Study in Sourcing Policy." *Journal of International Business Studies* 10: 9-20.

Burstein, D. 1988. *Yen: Japan's New Financial Empire and Its Threat to America*. New York: Simon and Schuster.

California Council for International Trade. 1990. *A Share in the Dream: Foreign Direct Investment in California*. San Francisco, CA: Niels Erich.

Campa, J. 1993. "Entry By Foreign Firms in the United States Under Exchange Rate Uncertainty." *The Review of Economies and Statistics* 75: 614-622.

Cantwell, J. 1989. *Technological Innovation and Multinational Corporations*. London: Blackwell.

______. 1994. *Transnational Corporations and Innovatory Activities.* London: Routledge.
Casson, M. 1979. *Alternatives to the Multinational Enterprise*. London: Macmillan.
______. 1986. "General Theories of the Multinational Enterprise: A Critical Examination." In *Multinationals: Theory and History*, edited by P. Hertner and G. Jones. Aldershot, VT: Gower.
Caves, R. 1971. "International Corporations: The Industrial Economies of Foreign Investment. *Economica* 38: 1-27.
______. 1974a. "Causes of Direct Investment: Foreign Firms' Share in Canadian and United Kingdom Manufacturing Industries." *Review of Economics and Statistics* 56: 279-293.
______. 1974b. "Multinational Firms, Competition, and Productivity in Host Country Markets." *Economica* 41: 176-192.
______. 1982. *Multinational Enterprise and Economic Analysis.* Cambridge, England: Cambridge University Press.
Caves, R. and S. Mehra. 1986. "Entry of Foreign Multinational into U.S. Manufacturing Industries." In *Competition in Global Industries*, edited by M. Porter. Boston: Harvard Business School.
______. 1989. "Exchange Rate Movements and Foreign Direct Investment in the United States." Pp. 199228 in *The Internationalization of U.S. Markets*, edited by D. B. Audretsch and M. P. Caudon. New York: New York University Press.
______. 1993. "Japanese Investment in the United States: Lessons for the Economic Analysis of Foreign Investment." *World Economy* 16(3): 279-300.
Chang, S. 1994. "Replication of Keiretsu in the United States: Transfer of Interorganizational Network Through Direct Investment." Paper presented at the 1994 Academy of International Business Conference, Boston, MA., November.
Chernotsky, H. 1987. "The American Connection: Motives for Japanese Foreign Direct Investment." *Columbia Journal of World Business* 22: 47-54.
Choate, P. 1990. *Agents of Influence*. New York: Alfred A. Knopf.
Clarke, M. 1986. *Revitalizing State Economies: A Review of State Economic Development Policies and Programs*. Washington, DC: National Governors' Association.
Coase, R. 1937. "The Nature of the Firm." *Economica* 4: 386-405.
Coase, R. and D. McFederidge. 1984. "International Technology and Theory of the Firm." *Journal of Industrial Economics* 32: 254-264.
Conway Data, Inc. 1991. *Site Selection and Industrial Development* (October): 952-1018.
______. 1993. *Site Selection and Industrial Development* (October): 1150-1186.
______. 1994. *Site Selection: The Global Magazine of Business Strategy* (October): 848-858.
______. 1995. *Site Selection: The Global Magazine of Business Strategy* (October): 726-800.
______. 1996. *Site Selection: The Global Magazine of Business Strategy* (October): 822-862.
Craig, T. 1995. "Location Decision and Implementation in Production and Non-Production FDI: The Case of Matsushita Electric Industrial Company." *Academy of Management Journal, Best Papers Proceedings* 167-171.

Cushman, D. 1988. "Exchange-Rate Uncertainty and Foreign Direct Investment in the United States." *Wellwirtschaftiches Archiv* 124: 322-336.

Davidson, W. 1980. "The Location of Foreign Direct Investment Activity: Country Characteristics and Experience Effects." *Journal of International Business Studies* 11: 9-22.

Denekamp, J. 1995. "Intangible Assets, Internalization and Foreign Direct Investment in Manufacturing." *Journal of International Business Studies* 26: 493-504.

Dewenter, K. 1995. "Do Exchange Rate Changes Drive Foreign Direct Investment?" *Journal of Business* 68: 405-433.

Dillon, K. 1989. *Japanese Investment in the United States*. Washington, DC: Foreign Service Institute, U.S. Department of State.

Dunning, J. ed. 1971. *The Multinational Enterprise*. London: Allen & Unwin.

______. 1980. "Towards an Eclectic Theory and International Production: Some Empirical Tests." *Journal of International Business Studies* 11: 9-31.

______. 1981a. *International Production and the Multinational Enterprise*. London: Allen & Unwin.

______. 1981b. "Explaining Outward Direct Investment of Developing Countries: In Support of the Eclectic Theory of International Production." In *Multinationals for Developing Countries*, edited by K. Kumar and M. G. McLeod. Lexington, MA: Lexington Books.

Dunning, J. and A. Rugman. 1985. "The Influence of Hymer's Dissertation on Theories of Foreign Direct Investment." *American Economic Review* 75: 228-232.

______. 1988a. *Explaining International Production*. London: Unwin and Hyman.

______. 1988b. "The Eclectic Paradigm of the International Production: A Restatement and Some Possible Extensions." *Journal of International Business Studies* 19: 1-31

______. 1995. "Reappraising the Eclectic Paradigm in An Age of Alliance Capitalism." *Journal of International Business Studies* 26: 461-489.

Dunning, J. and R. Narula. 1995. "The R&D Activities of Foreign Firms in the United States." *International Studies of Management and Organization* 25: 39-74.

Economic Policy Council, United Nations Association of the USA. 1991. *A Framework For International Direct Investment*. New York: United Nations Association of the USA.

Eisinger, P. 1988. *The Rise of the Entrepreneurial State: State and Local Economic Development Policy in the United States*. Madison, WI: University of Wisconsin Press.

Encarnation, D. 1986. "Cross-Investment: A Second Front of Economic Rivalry." Pp. 117-149 in *American Versus Japan*, edited by T. K. McCraw. Boston: Harvard Business School Press.

Erich, N. 1990. *A Share in the Dream: Foreign Direct Investment in California*. San Francisco: California Department of Commerce.

Ethier, W. 1986. "The Multinational Firm." *Quarterly Journal of Economics* 101(4): 805-833.

Faux, J. and W. Spriggs. 1991. *U.S. Jobs and the Mexico Trade Proposal*. Washington, DC: Economic Policy Institute.

Fierman, J. 1988. "The Selling of America." *Fortune* (May 23): 54-64.

Flowers, E. 1976. "Oligopolistic Reactions in European and Canadian Direct Investment in the United States." *Journal of International Business Studies* 7: 20-37.

Franko, L. 1971. *European Business Strategies in the United States.* Switzerland: Business International.

______. 1976. *The European Multinationals.* Stamford: Greylock Press.

______. 1983. *The Threat of Japanese Multinationals—How the West Can Respond.* New York: John Wiley.

Froot, K. and J. Stein. 1991. "Exchange Rates and Foreign Direct Investment: An Imperfect Capital Markets Approach." *The Quarterly Journal of Economics* 106: 1191-1217.

Fujita, M. 1995. "Small and Medium-Sized Transnational Corporations: Trends and Patterns of Foreign Direct Investment." *Small Business Economics* 7: 183-204.

Genther, P. and D. Dalton. 1990. *Japanese Investment in the United States.* Washington, DC: U.S. Department of Commerce.

Georgiou, G. and S. Weinhold. 1992. "Japanese Direct Investment in the U.S." *World Economy* 15: 761-778.

Gerlowski, D., H. Fung and D. Ford. 1994. "The Location of Foreign Direct Investment for U.S. Real Estate: An Empirical Analysis." *Land Economics* 70: 286-293.

Gilson, R. and M. Roe. 1992. "Understanding the Japanese Keiretsu: Overlaps Between Corporate Governance and Industrial Organization." Working Paper No. 97, Olin Program in Law and Economics, Sanford Law School.

Glick, R. 1990. "Japanese Capital Flows in the 1980s." *Economic Review: Federal Reserve Bank of San Francisco* 18-31.

Glickman, N. and D. Woodward. 1989. *The New Competitors: How Foreign Investors are Changing the U.S. Economy.* New York: Basic Books.

Goldberg, L. and C. Kolstad. 1995. "Foreign Direct Investment, Exchange Rate Variability and Demand Uncertainty." *International Economic Review* 36(4): 855-873.

Goldberg, M. 1972. "The Determinants of the U.S. Direct Investment in the E.E.C.: Comment." *The American Economic Review* LXII: 692-699.

Graham, E. 1974. *Investment and Growth in Mature Economies.* Oxford: Basil Blackwell and Mott.

______. 1978. "Transatlantic Investment by Multinational Firms: A Rivalistic Phenomenon." *Journal of Post Keynesian Economics* 1: 82-99.

Graham, E. and P. Krugman. 1995. *Foreign Direct Investment in the U.S.* Washington, DC: Institute for International Economics.

______. 1996. *Global Corporations and National Governments.* Washington, DC: Institute for International Economics.

Grosse, R. and L. Trevino. 1996. "Foreign Direct Investment in the United States: An Analysis by Country of Origin." *Journal of International Business Studies* 27: 139-155.

Gruber, W., D. Mehta and R. Vernon. 1976. "The R&D Factor in International Trade and International Investment of United States Industries." *Journal of Political Economy* 75: 43-55.

Hartman, D. 1984. "Tax Policy and Foreign Investment in the United States." *National Tax Journal* 37(4): 475-487.

He, X. and J. Li. 1996. "A Taxation Dilemma of Foreign Direct Investment in an Evolving Market Economy." *Multinational Business Review* 4: 36-49.

Heller, H. and E. Heller. 1974. *Japanese Investment in the United States*. New York: Prager Publishers.

Hennart, J. 1986. "What is Internalization?" *Weltwirtschaftliches Archiv* 122(4): 791-804.

______. 1989. "Can the New Forms of Investment Substitute for the Old Forms? A Transaction Cost Perspective." *Journal of International Business Studies* 20: 211-234.

Hennart, J. and Y. Park. 1994. "Japanese Manufacturing Investment in The United States." *Strategic Management Journal* 15: 419-436.

Hirsch, S. 1967. *Location of Industry and International Competitiveness*. Oxford: Clarendon Press.

Horaguchi, H. and B. Toyne. 1990. "Setting the Record Straight: Hymer, Internalization Theory and Transaction Cost Economics." *Journal of International Business Studies* 21: 487-494.

Horst, T. 1971. "The Theory of the Multinational Firm: Optimal Behavior under Different Tariff and Tax Rates." *Journal of Political Economy* 79: 1059-1072.

______. 1972. "Firm and Industry Determinants of the Decision of Invest Abroad: An Empirical Study." *Review of Economics and Statistics* 54: 258-266.

Hufbauer, G. and J. Schott. 1993. *NAFTA: An Assessment*. Washington, DC: Institute for International Economics.

Hultman, C. and R. McGee. 1988. "Factors Influencing Foreign Investment in the U.S., 1970-1985." *International Review of Economics and Business* XXXV: 1061-1067.

Hymer, S. 1976. *The International Operations of National Firms: A Study of Direct Foreign Investment*. Cambridge, MA: The MIT Press.

International Trade Administration, U.S. Department of Commerce. 1988. *International Direct Investment: Global Trends and the U.S. Role*. Washington, DC: U.S. Government Printing Office.

Janeba, E. 1995. "Corporate Income Tax Competition, Double Taxation Treaties, and Foreign Direct Investment." *Journal of Public Economics* 56: 311-325.

Johanson, J. and F. Wiedersheim-Paul. 1975. "The Internationalization of the Firm: Four Swedish Cases." *Journal of Management Studies* 12(3): 305-322.

Johanson, J. and J. Vahlne. 1990. "The Mechanism of Internationalization." *International Marketing Review* 7(4): 11-24.

Jones, K. 1994. *Export Restraint and the New Protectionism*. Ann Arbor, MI: University of Michigan Press.

Johnson, R. 1977. "Success and Failure of Japanese Subsidiaries in America." *Columbia Journal of World Business* 12: 30-37.

Kahley, W. 1987. "Direct Investment Activity of Foreign Firms." *Economic Review, Federal Reserve Bank of Atlanta* 72: 36-51.

Kim, S. and S. Kim. 1993. "Motives for Japanese Direct Investment in the United States." *Multinational Business Review* 6: 66-72.

Kim, S. and M. Nichols. 1995. "A Cross-Sectional Industry Analysis of Foreign Direct Investment in the U.S." *Multinational Business Review* 3: 68-73.

Kim, W. and E. Lyn. 1986. "Excess Market Value, the Multinational Corporation and Tobin's Q Ratio." *Journal of International Business Studies* 17: 119-125.

Kim, W. and R. Mauborgne. 1993. "Effectively Conceiving and Executing Multinationals' Worldwide Strategies." *Journal of International Business Studies* 24: 419-448.

Kimura, Y. 1989. "Firm-Specific Strategic Advantages and Foreign Direct Investment Behavior of Firm." *Journal of International Business Studies* 20: 296-314.

Kincaid, J. 1984. "The American Governors in International Affairs." *Publius* 14: 95-114.

Kindleberger, C. 1969. *American Business Abroad: Six Lectures on Direct Investment*. New Haven, CT: Yale University Press.

Kindleberger, C. and P. Lindert. 1978. *International Economics*. Homewood, IL: Richard D. Irwin, Inc.

Klein, L. 1983. *The Economics of Supply and Demand*. Baltimore, MD: Johns Hopkins University Press.

Klein, M. and E. Rosengren. 1994. "The Real Exchange Rate and Foreign Direct Investment in the United States: Relative Wealth vs. Relative Wage Effects." *Journal of International Economics* 36: 373-389.

Kline, J. 1982. *State Government Influence in U.S. International Economic Policy*. Lexington, MA: Lexington Books.

Knickerbocker, F. 1974. *Oligopolistic Reaction and Multinational Enterprise*. Cambridge, MA: Harvard University Press.

Koechlin, T. and M. Larudee. 1992. "The High Cost of NAFTA." *Challenge* 1-8.

Kogut, B. and V. Lander. 1992. "Knowledge of the Firm, Combinative Capabilities and the Replication of Technology." *Organization Science* 3(3): 383-397.

Kogut, B. and S. Chang. 1991. "Technological Capabilities and Japanese Foreign Direct Investment in the United States." *Review of Economics and Statistic* 75: 401-413.

Kogut, B. and U. Zander. 1993. "Knowledge of the Firm and the Evolutionary Theory of the Multinational Corporation." *Journal of International Business Studies* 24: 625-645.

Kojima, K. 1978. *Direct Foreign Investment: A Japanese Model of Multinational Business Operations*. New York: Prager.

Kojima, K. and T. Ozawa. 1985. "Towards a Theory of Industrial Restructuring and Dynamic Comparative Advantage." *Hitotsuboshi Journal of Economics* 26: 35-44.

Kotabe, M. 1993. "The Promotional Roles of the State Government & Japanese Manufacturing Direct Investment in the United States." *Journal of Business Research* 27(2): 131-146.

Kotabe, M. and G. Omura. 1989. "Sourcing Strategies of European & Japanese Multinationals: A Comparison." *Journal of International Business Studies* 20: 113-130.

Krug, J. and J. Daniels. 1994. "Latin American and Caribbean Direct Investment in the U.S." *Multinational Business Review* 2: 1-10.

Krugman, P. 1996. *Pop Internationalism*. Cambridge, MA: MIT Press.

Kujawa, D. 1986. *Japanese Multinationals in the United States: Case Studies*. New York: Prager.

Kwack, S. 1962. "A Model of U.S. Direct Investment Abroad: A Neoclassical Approach." *Western Economic Journal* X: 376-383.

Lall, S. and N. Siddharthan. 1982. "The Monopolistic Advantages of Multinationals: Lessons from Foreign Ivestment in the U.S." *The Economic Journal* 92: 668-683.

Lamfalussy, A. 1961. *Investment and Growth in Mature Economies*. Oxford: Basil Blackwell and Mott.

Lee, P. and W. Sullivan. 1995. "Considering Exchange Rate Movements in Economic Evaluation of Foreign Direct Investment." *Engineering Economist* 40: 171-199.

Leftwich, R. 1973. "Foreign Direct Investments in the United States 1962-1971." *Survey of Current Business* 29-40.

Lessard, D. 1979. "Transfer Prices, Taxes and Financial Markets: Implications of Internal Financial Transfers within the Multinational Firm." In *Economic Issues of Multinational Firms*, edited R. G. Hawkins. Greenwich, CT: JAI Press.

Lessard, J. 1995. "International Acquisition of U.S. Based Firms: Shareholder Wealth Implications." *American Business Review* 13: 50-57.

Levy, H. and M. Sarrat. 1970. "International Diversification in Investment Portfolios." *American Economic Review* LX: 668-675.

Lipsey, R. 1993. "Foreign Direct Investment in the United States: Changeover Three Decades." In *Foreign Direct Investment*, edited by K.A. Froot, Chicago: The University of Chicago Press.

Little, J. 1978. "Locational Decisions of Foreign Direct Investors in the U.S." *New England Economic Review, Federal Reserve Bank of Boston* 78: 43-63.

______. 1984. "The Industrial Composition of Foreign Direct Investment in the United States and Abroad: A Preliminary Look." *New England Economic Review, Federal Reserve Bank of Boston* 4: 38-48.

Loree, D. and S. Guisinger. 1995. "Policy and Non-Policy Determinants of U.S. Equity Foreign Direct Investment." *Journal of International Business Studies* 26: 281-299.

Love, J. 1995. "Knowledge, Market Failure and the Multinational Enterprise: A Theoretical Note." *Journal of International Business Studies* 26: 399-407.

Madura, J. and A. Whyte. 1990. "Diversification Benefits of Direct Foreign Investment." *Management International Review* 30: 73-85.

Magee, S. 1976. "Technology and the Appropriability Theory of the Multinational Corporation." In *The New International Order*, edited by Jagdish Bhagwati. Cambridge, MA: MIT Press.

______. 1977. "Multinational Corporations and the Industry Technology Cycle and Development." *Journal of World Trade Law* 11: 297-321.

Makin, J. 1988. "Japan's Investment in America: Is it a Threat?" *Challenge* 8-16.

______. 1989. "The Effects of Japanese Investment in the United States." Pp. 42-62 in *Japanese Investment in the United States: Should We Be Concerned?*, edited by K. Yamamura. Seattle: Society for Japanese Studies, University of Washington.

Mandell, S. and C. Killian. 1974. *An Analysis of Foreign Investment in Selected Areas of the United States: A Research Project on Behalf of the New England Regional Commission*. Boston, MA: The International Center of New England, Inc.

Markusen, J. 1995. "The Boundaries of Multinational Enterprises and The Theory of International Trade." *Journal of Economic Perspectives* 9: 169-189.

Martin, S. 1991. "Direct Foreign Investment in the United States." *Journal of Economic Behavior and Organization* 16(3): 283-293.

McCarthy, M. 1993. "Unlikely Sites: Why German Firms Chose the Carolinas to Build U.S. Plants." *Wall Street Journal* (May 4): 1.

McCulloch, R. 1993. "New Perspectives on Foreign Direct Investment." In *Foreign Direct Investment*, edited by K.A. Froot. Chicago, IL: University of Chicago Press.

McLure, C. 1992. "Substituting Consumption-Based Direct Taxation for Income Taxes as the International Norm." *National Tax Journal* 45: 145-154.

McWhirter, W. 1991. "The Bruising Battle Abroad." *Time* May 27, p. 42.

National Association of State Development Agencies. 1986. *Directors of Incentives for Business Investment and Development in the United States*. Washington, DC: The Urban Institute Press.

Neff, R. 1995. "Japan's New Identity." *Business Week* April 10, p. 108-114.

Neikirk, W. 1987. *Volcher, Portrait of the Money Man*. Chicago, IL: Congdon and Weed.

Negandhi, A. and M. Serapio. ed. 1992. "Japanese Direct Investments in the United States: Trends, Developments and Issues." In *Research in International Business and International Relations*. Greenwich, CT: JAI Press, Inc.

Nelson, R. and S. Winter. 1982. *An Evolutionary Theory of Economic Change*. Cambridge, MA: Harvard University Press.

Newman, R. and K. Rhee. 1990. "Midwest Auto Transplants: Japanese Investment Strategies & Policies." *Business Horizons*, 33: 63-69.

Noble, G. 1992. "Takeover or Makeover? Japanese Investment in America." *California Management Review* 34: 127-147.

OECD. 1989. *International Investment and Multinational Enterprise: Investment Incentives and Disincentives*. Paris: Organization for Economic Co-operation and Development.

______. 1995. *OECD Reviews of Foreign Direct Investment: United States*. Paris: Organization for Economic Co-Operation and Development.

Oman, C. 1984. *New Forms of International Investment in Developing Countries*. Paris: OECD.

Ondrich, J. and M. Wasylenko. 1993. *Foreign Direct Investment in the United States*. Kalamazoo, MI: W.E. Upjohn Institute for Employment Research.

Ott, M. 1989. "Is America Being Sold Out?" *Federal Reserve Bank of St. Louis Review* 47-64.

Oviatt, B. and P. McDougall. 1994. "Toward a Theory of International New Ventures." *Journal of International Business Studies* 25: 45-64.

Peat Marwick. 1991. *Survey of Foreign-Based Companies with Operations in Massachusetts*. Boston, MA: KPMG Peat Marwick.

Pechter, K. 1994. "The Many Attractions of Investing in the U.S." *International Business* 76-89.

Porcano, T. and C. Price. 1996. "The Effects of Government Tax and Non Tax Incentives on Foreign Direct Investment." *Multinational Business Review* 4: 9-19.

Porter, M. 1985. "Changing Patterns of International Competition." *California Management Review* XXVII: 9-38.

______. 1986. *Competition in Global Industries*. Boston: Harvard Business School Press.

Prestowitz, C. 1988. *Trading Places: How We Allowed the Japanese to Take the Lead*. New York: Basic Books.

Quickel, S. 1992. "Undaunted Venturers: Japan's Financiers Aim to Ride Out the Cash Crisis." *International Business* 5: 52-55.

Ray, E. 1988. "The Determinants of Foreign Direct Investment in the United States, 1979-85." Pp. 53-77 in *Trade Policies for International Competitiveness*, edited by R. E. Feenstra. Chicago: University of Chicago Press.

______. 1991. "Foreign Takeovers & New Investments in the United States." *Contemporary Policy Issues* 9: 59-71.

Rivoli, P. and E. Salorio. 1996. "Foreign Direct Investment and Investment Under Uncertainty." *Journal of International Business Studies* 27: 335-357.

Root, F. and A. Ahmed. 1978. "The Influence of Policy Instruments on Manufacturing Direct Foreign Investment in Developing Countries." *Journal of International Business Studies* 9: 81-94.

Rogers, C. 1995. "Global Information Review." *Indiana Business Review* 70: 1.

Rolfe R., D. Ricks, M. Pointer and M. McCarthy. 1993. "Determinants of FDI Incentive Preferences of MNEs." *Journal of International Business Studies* 24: 335-355.

Rugman, A. 1980. "Internationalization as a General Theory of Foreign Direct Investment: A Re-appraisal of the Literature" *Weltwirtschaftliches Archiv* 116: 368-369.

______. 1981. *Inside the Multinationals: The Economies of International Markets*. London: Croom Helm.

______. 1986. "New Theory of the Multinational Enterprise: An Assessment of Internationalization Theory." *Bulletin of Economic Research* 38(2): 101-118.

Rutter, J. 1991. "Trends and Patterns in Foreign Direct Investment in the United States." In *Economics and Statistics Administration, Office of Chief Economist, Foreign Direct Investment in the U.S.: Review and Analysis of Current Developments*. Washington, DC: U.S. Government Printing Office.

Salvatore, D. 1991. "Trade Protection & Foreign Direct Investment in the United States." *The Annals of the American Academy of Political & Social Science* 91-105.

Sametz, A. and J. Backman. 1974. "Why Foreign Multinationals Invest in the United States." *Challenge* 43-47.

Sapsford, J. and R. Steiner. 1995. "Huge Japanese Merger Could Help Revitalize the Financial Sector." *Wall Street Journal* pp. 1, 8.

Sarathy, R. 1985. "Japanese Trading Companies: Can They Be Copied?" *Journal of International Business Studies* 15: 101-119.

Scaperlanda, A. 1967. "The EEC and U.S. Foreign Investment: Some Empirical Evidence." *Economic Journal* 77: 22-25.

Scarperlanda, A. and L. Mauer. 1969. "The Determinants of U.S. Direct Investment in the E.E.C."*American Economic Review* LIX: 558-568.

Schott, J. 1994. *The Uruguay Round: An Assessmen*. Washington, DC: Institute for International Economics.

Servan-Schreiber, J. J. 1967. *Le Defi Americain*. Paris: Denoel.

Smothers, N. 1990. "Patterns of Japanese Strategy: Strategic Combinations of Strategies." *Strategic Management Journal* 11: 521-533.

Sokoya, S. and K. Tillery. 1992. "Motives of Foreign MNCs Investing in the United States and Effect of Company Characteristics." *The International Executive* 34: 65-80.

Stevens, G. 1973. "The Multinational Firm and the Determinants of Investment." Discussion Paper No. 29, Division of International Finance, Board of Governors of the Federal Reserve System.

Stopford, J., S. Strange and J. Henley. 1992. *Rival States, Rival Firms: Competition for World Market Shares*. Cambridge: Cambridge University Press.

Streeten, P. 1974. "The Theory of Development Policy." In *Economic Analysis and the Multinational Enterprise*, edited by J. Dunning. New York: Praeger.

Strozier, R. 1988. "Is the U.S. Up for Sale? A Growing Debate." *World* 7-8.

Swamidass, P. and M. Kotabe. 1993. "Component Sourcing Strategies of Multinationals: An Empirical Study of European & Japanese Multinationals." *Journal of International Business Studies* 24: 81-99.

Swenson, D. 1994. "The Impact of U.S. Tax Reform on Foreign Direct Investment in the United States." *Journal of Public Economics* 54: 243-266.

Takaoka, H. 1991. "Present Status and Impact of Japan's Investment in Manufacturing Industries in the United States." *EXIM Review* 11: 57-83.

The Boston Globe. 1992. "Foreign Investors Like New England." May 29, p. 26.

The Wall Street Journal. 1977. "Top Soy Sauce Brewer in Japan Shows How to Crack U.S. Market." September 16, p. 1.

______. 1979a. "Foreign Firms Step Up Takeovers in the U.S. and Worry in Rising." April 20, p. 1.

______. 1979b. "Weak Dollar, Stocks Spur Foreigners to Seek Acquisitions in the U.S." August 21, p. 1.

Thornton, E. 1992. "How Japan Got Burned in the U.S.A." *Fortune* (June 15), pp. 114-116.

Thurow, L. 1992. *Head to Head*. New York. William Morrow & Company.

Tolchin, M. and S. Tolchin. 1988. *Buying Into America: How Foreign Money is Changing the Face of Our Nation*. New York: Time Books.

Tsurumi, Y. 1976. *The Japanese Are Coming: A Multinational Spread of Japanese Firms*. Cambridge: Bailinger.

______. 1977. *Multinational Management: Business Strategy and Government Policy.* Cambridge, MA: Ballinger Publishing Company.

Tzu, S. 1983. *The Art of War.* New York. Dell Publishing.

United Nations Centre on Trade and Development. 1995. *World Investment Report: 1995*. New York: United Nations.

United Nations Centre on Transnational Corporations. 1992. *The Determinants of Foreign Direct Investment: A Survey of the Evidence*. New York: United Nations.

U.S. Conference Board. 1973. *The Multinational Corporation: Studies on U.S. Foreign Investment*. Washington, DC: U.S. Government Printing Office.

U.S. Department of Commerce. 1976. *Foreign Direct Investment in the United States: Report of the Secretary of Commerce to the Congress in Compliance with the Foreign Investment Study Act of 1974. Vol. 5, Appendix G*. U.S. Government Printing Office.

______. 1987. *Foreign Direct Investments in the United States: 1987 Benchmark Survey, Final Results*. Washington, DC: Bureau of Economic Analysis.

______. 1991. *Foreign Direct Investments in the United States: Review and Analysis of Current Developments*. Washington, DC: Office of the Chief Economist, Economics & Statistics Administration.

______. 1993a. *Foreign Direct Investment in the United States: An Update*. Washington, DC: Office of the Chief Economist, Economics and Statistics Administration.

______. 1974. *Survey of Current Business*, August, pp. 7-9.

______. 1978. *Survey of Current Business*, August, pp. 39-52.

______. 1982. *Survey of Current Business*, August, pp. 30-41.

______. 1988. *Survey of Current Business*, August, pp. 69-83.

______. 1992. *Survey of Current Business*, August, pp. 87-115.

______. 1993b. *Survey of Current Business*, July, pp. 59-87.

______. 1994. *Survey of Current Business*, June, pp. 154-186.

______. 1995. *Survey of Current Business*, August, pp. 53-87.

______. 1996. *Survey of Current Business*, July, pp. 102-130.

______. 1997. *Survey of Current Business*, September, pp. 75-118.

Vaitsos, C. 1974. *Intercountry Income Distribution and Transnational Enterprises*. Oxford: Clarendon Press.

Vernon, R. 1966. "International Investment and International Trade in the Product Cycle." *Quarterly Journal of Economics* 83(1): 190-207.

______. 1970. *Sovereignty at Bay*. New York: Basic Books.

______. 1974. "The Location of Economic Activity." In *Economic Analysis and the Multinational Enterprise*, edited by J. H. Dunning. London: Allen and Unwin.

______. 1984. "Competition Policy Toward Multinational Corporations." *American Economic Review* 74: 276-282.

Wassman, U. and K. Yamamura. 1989. "Do Japanese Firms Behave Differently? The Effects of Keiretsu in the United States" In *Japanese Investment in the United States: Should We Be Concerned?*, edited by K. Yamamura. Seattle: Society for Japanese Studies, University of Washington.

Webley, S. 1974. *Foreign Direct Investment in the United States: Opportunities and Impediments*. London: British-North American Committee.

Wells, L. 1972. *The Product Life Cycle in International Trade*. Boston, MA: Division of Research, Graduate School of Business Administration, Harvard University.

Wilkins, M. 1989. *The History of Foreign Investment in the United States*. Cambridge, MA: Harvard University Press.

______. 1990. "Japanese Multinationals in the United States: Continuity and Change, 1879-1990." *Business History Review* 64: 585-629.

Williamson, O. E. 1964. *The Economics of Discretionary Behavior: Managerial Objectives in the Theory of the Firm*. Englewood Cliffs, NJ: Prentice Hall.

______. 1970. *Corporate Control and Business Behavior*. Englewood Cliffs, NJ: Prentice Hall.

______. 1975. *Markets and Hierarchies: Analysis and Antitrust Implications*. New York: Free Press.

______. 1981. *The Modern Corporation: Origins, Evolution, Attributes: Multinationals as Mutual Invaders*. London: Croom Helm.

World Trade Organization. 1996. *WTO Press Release, Issued by the Information and Media Relations Division of the World Trade Organization*, October 16, Geneva, Switzerland.

Yamamura, K. ed. 1989. *Japanese Investment in the United States: Should We Be Concerned?* Seattle: Society for Japanese Studies, University of Washington.

Yoffie, D. 1993. "Foreign Direct Investment in Semiconductors." Pp. 197-230 in *Foreign Direct Investment*, edited by K. Froot. Chicago, IL: University of Chicago Press.

Yoshida, M. 1975. "Japanese Foreign Direct Investment." Pp. 248-272 in *The Japanese Economy in International Perspective*, edited by I. Frank. Baltimore, MD: The Johns Hopkins University Press.

______. 1987a. "Macro-Micro Analysis of Japanese Manufacturing Investment in the United States." *Management International Review* 27: 19-32.

______. 1987b. *Japanese Direct Manufacturing Investment in the United States*. New York: Praeger.

Young, K. 1988. "The Effects of Taxes and Rates of Return on Foreign Direct Investments in the United States." *National Tax Journal* 41: 109-121.

Yu, C. and K. Ito. 1988. "Oligopolistic Reaction and Foreign Direct Investment: The Case of the U.S. Tire and Textile Industries." *Journal of International Business Studies* 19: 449-460.

______. 1990. "The Experience Effect and Foreign Direct Investment." *Weltwirtschaftliches Archiv* 126(3): 561-580.

Index